Jianjun Xiao, Jack Travis, Peter Royl, Gottfried Necker
Anatoly Svishchev, Thomas Jordan

GASFLOW-MPI: A Scalable Computational Fluid Dynamics Code for Gases, Aerosols and Combustion

Volume 1: Theory and Computational Model (Revision 1.0)

Karlsruhe Institute of Technology
KIT SCIENTIFIC REPORTS 7710

GASFLOW-MPI:
A Scalable Computational Fluid Dynamics Code for Gases, Aerosols and Combustion

Volume 1: Theory and Computational Model (Revision 1.0)

by
Jianjun Xiao, Jack Travis, Peter Royl, Gottfried Necker
Anatoly Svishchev, Thomas Jordan

KIT Scientific Publishing

Report-Nr. KIT-SR 7710

Impressum

KIT Scientific Publishing

Karlsruher Institut für Technologie (KIT)
KIT Scientific Publishing
Straße am Forum 2
D-76131 Karlsruhe

KIT Scientific Publishing is a registered trademark of Karlsruhe Institute of Technology. Reprint using the book cover is not allowed.

www.ksp.kit.edu

Print on Demand 2016

ISBN 978-3-7315-0448-1 (Vol. 1)
ISBN 978-3-7315-0449-8 (Vol. 2)
ISBN 978-3-7315-0447-4 (Set)
DOI: 10.5445/KSP/1000050393

Executive Summary

Karlsruhe Institute of Technology (KIT) is developing the parallel computational fluid dynamics code GASFLOW-MPI as a best-estimate tool for predicting transport, mixing, and combustion of hydrogen and other gases in nuclear reactor containments and other facility buildings. The code can model geometrically complex facilities with multiple compartments and internal structures. It can simulate the effects of two-phase dynamics with the homogeneous equilibrium model (HEM), two-phase heat transfer to walls and internal structures, chemical kinetics, catalytic recombiners, and fluid turbulence. An analysis with the GASFLOW-MPI code will result in the complete fluid dynamics description of gas species and discrete particle distribution and pressure, and temperature loadings on the walls and internal structures participating in an event.

GASFLOW sequential version has been used to calculate the distribution and control of hydrogen and noxious gases in complicated nuclear containment and confinement buildings and in nonnuclear facilities. It has been applied to situations involving transporting and distributing combustible gas mixtures. It has been used to study gas behavior in complicated containment systems with low-speed buoyancy-driven flows, with diffusion-dominated flows, and during deflagrations. The effects of controlling such mixtures by safety systems can be analyzed.

GASFLOW-MPI is a finite-volume code based on proven computational fluid dynamics methodology that solves the compressible Navier-Stokes equations for three-dimensional volumes in Cartesian or cylindrical coordinates. Wall shear stress models are provided for bulk laminar and turbulent flow. GASFLOW-MPI has transport equations for multiple gas species and one for internal energy. The two turbulence models available in GASFLOW-MPI are the algebraic and κ–ε model which provide zero- and two-transport-equation models that determine turbulent velocity and length scales needed to compute the turbulent viscosity. Terms for turbulent diffusion of different species are included in the mass and internal energy equations.

Heat conduction within walls and structures is one dimensional. Heat and mass transport to walls and structures is based on a modified Reynolds-Chilton-Colburn analogy, which accounts for increased heat transfer and condensation when the mass fraction of steam becomes a relatively large fraction of the mass of the gas mixture. Vaporization of fluid films is included with an inhibiting function as water vapor concentrations in fluid volumes adjacent to structures increase. Two-phase dynamics can occur in the fluid mixture volumes according to a classical homogeneous equilibrium model.

Chemical energy of combustion involving hydrogen provides a source of energy within the gaseous regions. A one-step global chemical kinetics model based on a modified Arrhenius law accounts for local hydrogen and oxygen concentrations. Models based on combustion progress variable transport equation have been also developed in the GASFLOW-MPI code. Hydrogen is ignited using a generalized ignitor model that represents both spark- and glow-plug-type designs. A catalytic hydrogen combination with oxygen is modeled using data from both the Nonproliferation and International Security division (NIS) and Siemens recombiner box designs.

The aerosol model comprises the following models: Lagrangian discrete particle transport, stochastic turbulent particle diffusion, particle deposition, particle entrainment, and particle cloud. These models incorporate the physics of particle behavior to model discrete particle phenomena and allow the code user to track the transport, deposition, and entrainment of discrete particles as well as clouds of particles.

In GASFLOW-MPI, the computational domain is discretized by a mesh of rectangular parallelepiped cells in either Cartesian or cylindrical geometry where primary hydrodynamic variables are cell-face-centered normal velocity and cell-centered density, internal energy, and pressure. A linearized Arbitrary-Lagrangian-Eulerian method is used for approximating the solution to the mass, momentum, and energy conservation equations.

The code version described in this manual is designated GASFLOW-MPI 1.0. In the 1980s, the name Hydrogen Mixing Studies, or HMS, was applied to any of a series of codes developed to solve special problems in HMS using a common theoretical basis. The latest version of HMS (HMS-93, for the year 1993, also known as HMS 1.0 for the first integrated version) integrated the best features of all the older versions into a single software package. This work was sponsored by the US Nuclear Regulatory Commission (NRC) as a best-estimate tool for nuclear containment analyses involving hydrogen and cooling issues. HMS 1.0 is the initial version of a larger code package called GASFLOW, which is supported by the US Department of Energy (DOE) to address various nuclear and nonnuclear facility safety issues. HMS 1.0 for the NRC is the same as GASFLOW 1.0 for the DOE.

Previous versions of HMS were applied to the following facilities and standard problems:

- EPRI/HDR International Standard Problems.
- Sandia FLAME and VGES Facilities.
- Nevada Hydrogen Tests.
- NRC Containment Loads Working Group Standard Problems.
- HCOG 1/4 Scale Test Facility.
- CSNI Hydrogen Distribution Benchmark Problems.
- Hydrogen Rule for Large Dry Containments.
- PHDR Large-Scale Hydrogen Mixing Experiment.
- PHDR Fire Experiments.

GASFLOW 2.0 has been extended beyond GASFLOW 1.0 with the following developments:

- Independent multiblock computational domains.
- Independent multiblocks connected on external boundaries by a ventilation system.
- Implemention of a fraction area treatment to model flow areas smaller than a cell face area.
- Accurate internal energy as a function of temperature to 4th degree polynomials.
- Gas properties library of thermochemical and transport extended to 30 species.
- Homogeneous equilibrium model for fluid mixture.
- Droplet depletion or "rainout".
- Two-phase heat and mass transfer to structural components.
- Both spark- and glow-plug ignitor models.

- Empirical hydrogen combustion limits.
- Hydrogen recombiner models.
- Transport, deposition, and entrainment of discrete particles.

GASFLOW-MPI 1.0 has been extended beyond the GASFLOW serial version 3.5 using the paradigms of Message Passing Interface (MPI) and domain decomposition. The data structure, parallel linear solvers and preconditioners in Portable Extensible Toolkit for Scientific Computing (PETSc) has been employed. GASFLOW-MPI 1.0 has been parallelized based on GASFLOW serial version 3.5 with major changes in the following:

- The data structure in GASFLOW serial code has been completely replaced by using the distributed arrays in PETSc library.
- The Precontioned Conjugate Residual method used in GASFLOW serial version has been replaced by the parallel preconditioners and linear solvers in PETSc library.
- Multiblock computational domains and multiblocks connected on external boundaries by a ventilation system is not supported in the current GASFLOW-MPI release. In order to keep the backward compitability, the block number in input variable arrays in ingf file, such as gasdef(7,*), mobs(7,*) and walls(7,*), were not removed. However, currently it must be always 1 for GASFLOW-MPI applications.
- Transport, deposition, and entrainment of discrete particles are currently not supported. These features will be implemented in the next release of GASFLOW-MPI.
- Only algebraic and κ-ε turbulence models are currently supported. More turbulence models will be implemented in future release of GASFLOW-MPI.
- Models for turbulent combustion have been extended in GASFLOW-MPI.
- Post-processing: pscan and graphic library, cgs, will not be used as post-processing tool in GASFLOW-MPI. Data format for GMV, Opendx, AVS and VISART are not suppoted. Instead, we provide Python tools, pyscan and create3D, for visualization purposes. The data can be read by the most popular 3D visualization tools, such as Paraview, Visit, Tecplot and Ensight.

Acknowledgements

We would like to thank all GASFLOW users for their bug reports and suggestions during all these years. We thank the financial supports from U.S. NRC and U.S. DOE for the development of GASFLOW serial version in 1980s and 1990s. We also would like to thank J. W. Spore, K. L. Lam, C. Müller, B. D. Nichols, T. L. Wilson of the Los Alamos National Laboratory for all the previous work that has been accomplished in the GASFLOW serial version. We greatly acknowledge B. Smith and the PETSc development team of Argonne National Laboratory for their support and help during the development of parallel version GASFLOW-MPI.

Contents

Nomenclature

a	Gordon and McBride coefficient for specific heat polynomial, CHEMKIN coefficients for transport properties, and coefficient for internal energy polynomial
A	Hamaker constant and generalized polynomial coefficients for internal energy approximation
$\boldsymbol{A}$	Fractional flow-area vector
Ar	Archimedes number
A_s	Wall or structural surface area
Area	Area of fluid cell in contact with GRS recombiner foil
b	Coefficient for internal energy polynomial
$\mathbf{b}$	Velocity of control surface
c	Coefficient for internal energy polynomial, water saturation curve coefficients, and species concentration when written with a component subscript
C	Relaxation coefficient for phase-change model (HEM) or droplet rainout model (s^{-1})
C_c	Energy of combustion for hydrogen
C_D	Fluid drag coefficient
$\mathbf{C}_D$	Vector structural drag coefficient
C_f	Frequency (or pre-exponential) factor
C_p	Specific heat at constant pressure
C_v	Specific heat at constant volume
d	Coefficient for internal energy polynomial
d	Particle (sphere) diameter (cm)
d_p	Particle diameter (cm)
$\mathbf{D}_d$	Structural drag vector
$D_{\alpha\to mix}$	Molecular diffusion coefficient of species α into the mixture
$D_{\alpha\to\beta}$	Binary molecular diffusion coefficient of species α into species β
e	Coefficient for internal energy polynomial
e	Coefficient of restitution
E	Activation energy
E_i	Potential energy well depth on impact (ergs)
E_r	Potential energy well depth on rebound (ergs)
f_s	Static friction coefficient
$\mathbf{g}$	Acceleration of gravity (cm/s^2)
h	Apparent particle penetration depth (cm)
h_d	Structural mass-transfer coefficient
h_d^*	Corrected structural mass-transfer coefficient for effect of water vapor
h_s	Structural heat-transfer coefficient
h_s^*	Corrected structural heat-transfer coefficient for effect of water vapor

h_{α}	Enthalpy for species α
h_2	Hydrogen
h_2o	Water vapor
H	Heaviside function
i	x- (or r-) direction index
I	Specific internal energy
$I_{reference}$	Specific internal energy at reference temperature (298.15 K)
j	y- (or θ-) direction index
J_{α}	Mass flux vector for species α
k	z-direction index
k(T)	Chemical rate constant
K	Bulk mechanical property: = $(1-N^2)/Y$ (cm^2/dyne)
KE_i	Kinetic energy from incident normal velocity (ergs)
KE_r	Kinetic energy from rebound normal velocity (ergs)
l	Turbulence length scale
m	Mass
m	Particle mass (g)
$\dot{m}_s$	Mass transfer rate to and from structural surface
M	Molecular weight
n	Steam mole fraction
n	Unit normal vector
n_2	Nitrogen
o_2	Oxygen
p	Pressure
Pr	Prandtl number
Q	Volumetric flow rate into recombiner
q	Energy flux vector
$q_{\rho,h2o}$	Energy lost or gained by the structure due to phase change
q_s	Convected energy to structural surface
R	Radial coordinate
R	Flux ratio
R	Radius of curvature (cm)
R_{α}	Gas constant for species α
$\dot{R}_{h2}$	Recombination reaction rate
$\leftarrow$	Universal gas constant
S	Fluid cell surface area
S^*	Moving control surface
Sc	Schmidt number
S_I	Energy source or sink
S_m	Momentum source or sink
S_{Φ}	Arbitrary source term

S_ρ	Mass source or sink resulting from phase change involving structures
$S_{\rho,\alpha}$	Mass source or sink resulting from chemical reactions and HEM
t	Time
T	Temperature
u	x- (or r-) direction velocity component
u	Fluid velocity vector
$\mathbf{u_g}$	Grid velocity vector
u_y	Velocity component tangent to and a distance *y* from wall (cm/s)
u_*	Shear velocity (cm/s)
U	Velocity into recombiner
$\mathbf{U_g}$	Gas velocity (cm/s)
$\mathbf{U_{gcp}}$	Gas velocity at center of particle (cm/s)
$\mathbf{U_{gpu}}$	Minimum particle pickup velocity predicted by the semi-empirical model (cm/s)
$\mathbf{U_{gpuo}}$	Minimum particle pickup velocity predicted by the single particle model (cm/s)
$\mathbf{U_i}$	Incident normal velocity (cm/s)
$\mathbf{U_i^*}$	Critical normal rebound velocity (cm/s)
$\mathbf{U_p}$	Particle velocity (cm/s)
$\mathbf{U_r}$	Rebound normal velocity (cm/s)
$\mathbf{U_t}$	Tangential velocity (cm/s)
v	y- (or θ-) direction velocity component
V	Volume
V^*	Moving control volume
w	z-direction velocity component
x	Position vector
x	Cartesian coordinate
x_α	Mass fraction of species α
y	Cartesian coordinate
y	Distance normal to surface and perpendicular to flow direction (cm)
y_c	Distance from wall to cell center
Y	Particle Young's modulus (dyne/cm^2)
Y_α	Molar or volume fraction for species α
z	Cartesian coordinate
z_0	Separation distance of sphere and substrate (cm)
α	Coefficient of thermal expansion
β	Wall thermal diffusivity
d_l	Boundary layer thickness (cm)
Δt	Time step
Δx	Mesh cell size for heat conduction grid
δx	Mesh cell size for fluid cells in x- or r- coordinate direction
δy	Mesh cell size for heat cells in y- or θ- coordinate direction

δz	Mesh cell size for heat cells in z-coordinate direction
η	Molecular viscosity (g/cm s) and recombiner efficiency
ε	Dissipation of turbulent kinetic energy and time-step parameter
κ	Turbulent kinetic energy
ν	Kinematic viscosity (cm^2/s)
N	Poisson ratio
λ	Second coefficient of molecular viscosity
μ	First coefficient of molecular viscosity
ν	Molecular kinematic viscosity
ρ	Density
r_g	Gas density (g/cm^3)
r_p	Particle density (g/cm^3)
σ	Turbulence coefficient
τ	Viscous stress tensor
τ_S	Wall or structural shear stress
θ	Azimuthal coordinate and gas mixture volume or void fraction
ω	Reaction rate
ϕ	Molecular thermal conductivity
Φ	Arbitrary scalar or vector function
ϕ_T	Rate factor
Ψ	General molecular transport coefficient and turbulence variable
Θ_m	Mass-transfer correction factor
Θ_T	Heat-transfer correction factor

Superscripts

A	Lagrangian Phase A computational level
B	Lagrangian Phase B computational level
n	Old time level
n+1	New time level
*	Moving control surface or volume and corrected heat and mass Transfer coefficients for water vapor effect
o	Reference value
	Thermodynamic or microscopic density

Subscripts

b	Bulk fluid property
c	Cell-centered value and combustion convection
	Related to convective heat-transfer condensation/vaporization
	Related to two-phase interaction combustion
	Related to combustion
d	Related to mass-transfer coefficient
f	Cell face
g	Gas
h_2	Hydrogen component
h_2o	Water vapor component
h_2ol	Water liquid component
i	x- (or r-) direction index
in	Recombiner inlet
I	Related to the energy conservation equation
j	y- (or θ-) direction index
k	z-direction index
m	Momentum control volume, or related to the momentum conservation equations
max	Maximum droplet loading of the atmosphere mixture
	Refers to fluid mixture
out	Recombiner outlet
o_2	Oxygen component
0	Reference
	Rainout
	Referees to droplet rainout model
ref	Reference
s	Structure surface
	saturation or sat
	Saturation condition for structural surface or mixture temperature
t	Turbulent and total or apparent
total	Total heat flux to structures, convection plus phase change
w	Wall-temperature distribution in structures
x	Vector component in x- or r-coordinate direction
y	Vector component in y- or θ-coordinate direction
z	Vector component in z-coordinate direction
α	Species
ε	Related to the turbulence coefficients
κ	Related to the turbulence coefficients
μ	Related to the turbulence coefficients
ρ	Related to the mass conservation equations

1 Introduction

1.1 Background

Combustion in nuclear facility buildings, such as light-water reactor (LWR) containments, can cause high pressures or temperatures that can, in turn, damage the containment or affect important safety-related equipment. After the Three Mile Island accident (a severe, or degraded-core, accident), it was found that significant quantities of hydrogen had been generated from the chemical reaction between the zirconium cladding (the thin protective covering of the nuclear fuel) and the water vapor. When released into the containment, this hydrogen burned by one or more combustion modes and threatened the containment integrity, internal structures, and safety-related equipment.

Modeling the geometries of containment buildings is difficult. One example is the Heiss Dampf Reactor (HDR) containment near Frankfurt, West Germany, which is shown in Figure 1-1. The HDR building is 60 m high and 20 m in diameter. It contains two stairwells, an elevator shaft, several vertical open hatchways, and ~70 rooms. This particular containment has roughly 11,300 m^3 of free volume, or approximately one-sixth to one-eight the free volume of a typical United States (US) or German Convoi-type pressurized water reactor (PWR) containment.

The German Nuclear Utility Industry, Siemens, the US Department of Energy (DOE), and US Nuclear Regulatory Commission (NRC) are supporting research at the Forschungszentrum Karlsruhe and Los Alamos National Laboratory to develop GASFLOW to evaluate design basis and beyond design basis severe accidents in nuclear reactor containments as well as in other nuclear facility buildings and the consequences on safety-related equipment and the containment structure itself. Current research coordinates model development with validation using experiments performed in Germany, the European Communities, and the US. We will describe the Los Alamos **I**mplicit **C**ontinuous-Fluid **E**ulerian **A**rbitrary-**L**agrangian-**E**ularian numerical methodology (ICE'd ALE) field-model approach in the report.

1.2 Computational Methodology

This report documents the theoretical and computational aspects of GASFLOW-MPI, a scalable finite-volume computer code for solving transient, three-dimensional (3D), compressible, Navier-Stokes equations for multiple gas species.

The code is designed to be a best-estimate tool for predicting the transport, mixing, and combustion of hydrogen gas in nuclear reactor containments and other nuclear facility buildings and structures. GASFLOW-MPI is based on the governing physical laws and modeling assumptions that are described in Chapter 0. In Chapter 3, we describe the linearized ICE'd ALE that provides the basis of the computational method used to integrate the equations in time and space. Briefly, each computational step is divided into three phases.

(1) An explicit Lagrangian phase computes changes in material volume, density, velocity, and internal energy caused by pressure gradients, combustion ignited with a generalized ignitor model, condensation and vaporization within the assumptions of the two-phase homogeneous equilibrium model, a structural two-phase heat transfer, catalytic recombiner hydrogen mitigation, structural heat conduction, and turbulence.

(2) An implicit Lagrangian phase calculates pressures at the advanced time level by solving simultaneously for pressure, density, velocities, and internal energy.

(3) A rezone phase computes the mass, momentum, and energy exchange between Eulerian cells that has occurred in the Lagrangian phase and repartitions or rezones these variables onto the original mesh.

The computational time step is completed with the evaluation of turbulence quantities, such as kinetic energy, dissipation and viscosity, which are calculated explicitly; the discrete particle transport, deposition, and entrainment governing equations are solved independently of the fluid flow equations; and a global analysis of the stability time step is performed at the end of each time step.

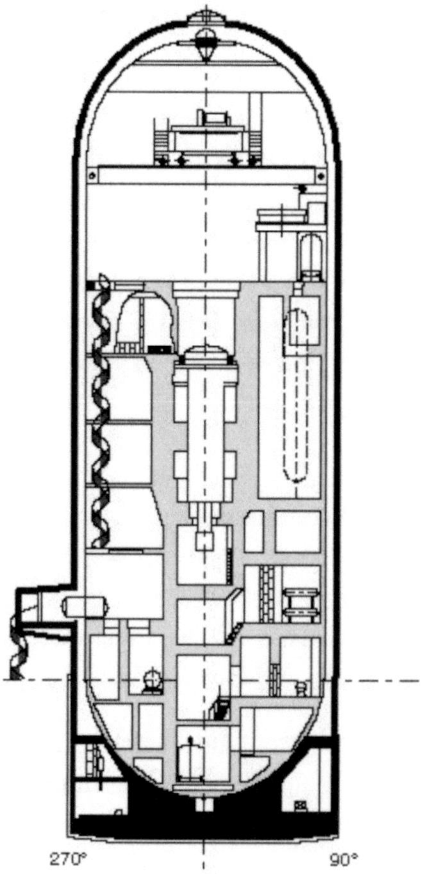

Figure 1-1 Cross section of the Heiss Dampf Reactor near Frankfurt, Germany

2 Mathematical and Physical Models

The equations of motion for a compressible fluid are derived from the physical laws that require the conservation of mass, momentum, and energy. The equations of change, which are presented in this section, relate the dynamics of the fluid to temporal and spatial influences such as viscous stress, body force, turbulence, structural resistance, heat transfer, phase change, and combustion. This includes relations for the transport of individual gas species. An equation of state is included to relate pressure (p) to density (ρ), mixture temperature (T), and volume fraction occupied by the gaseous mixture (θ).

As suggested by the "ALE" name, GASFLOW-MPI uses both the Lagrangian and Eulerian methodologies. The Lagrangian (or material) specification considers specific elements of matter and describes the motion as functions of space (**x**) and time (t). This approach is useful because the conservation laws refer to specific parcels of matter. However, the Eulerian (or spatial) viewpoint is often more convenient because it describes flow in terms of volumes fixed in space. Because the computational method used to model the flow is facilitated by dividing the problem domain into parallelepiped Eulerian volumes (cells) in either Cartesian or cylindrical geometry, it is natural to present the continuous equations in integral form (Ref. 2-1. and Ref. 2-2). This makes it easier to see how the integration of continuous volume and surface integrals presented in this section is approximated by the discrete or finite-volume equations given in Chapter 1.

2.1 The Generalized Conservation Equation

The conservation of any arbitrary extensive variable (for example, mass, momentum, or energy) is

$$\frac{\partial}{\partial t}\int_V \Phi dV = \int_V S_\Phi dV \quad , \tag{2-1}$$

where $\Phi(\mathbf{x}, t)$ is any continuously summable function, V is a material volume, and S_Φ is a source term. We use the Reynolds Transport Theorem, which may be expressed as

$$\int_V \frac{\partial \Phi}{\partial t} dV = \oint_S \Phi(\mathbf{u}\cdot\mathbf{A})dS + \int_V S_\Phi dV \quad , \tag{2-2}$$

where $\mathbf{u}(\mathbf{x}, t)$ is the fluid velocity and $\mathbf{A}$ and dS are the outward normal fractional area vector and differential area, respectively, of material surface S bounding V. We introduce the fractional area methodology of the FAVOR (Fractional Area Volume Obstacle Representation) algorithm due to Hirt (Ref. 2-3) to more easily and accurately model variable flow areas involved in complex geometrical representations. Applying the transport theorem to an arbitrary control volume V^* (Ref. 2-2) (not necessarily a material volume) enclosed by surface S^* gives

$$\frac{\partial}{\partial t}\int_{V^*} \Phi dV = \int_{V^*} \frac{\partial \Phi}{\partial t} dV + \oint_{S^*} \Phi(\mathbf{b}\cdot\mathbf{A})dS + \int_{V^*} S_\Phi dV \quad , \tag{2-3}$$

and add in Equation (2-3) where **b** is the velocity of the contour surface S'. If V^* is chosen to be instantaneously coincident with *V*, Equations (2-2) and (2-3) may be combined to give

$$\frac{\partial}{\partial t}\int_{V^*}\Phi dV=\oint_{S^*}\Phi(\mathbf{b}-\mathbf{u})\cdot\mathbf{A}dS+\int_{V^*}S_\Phi dV\,, \tag{2-4}$$

the integral form of the generalized conservation law. This is the basic kinematic relation used in the following three subsections and states that the time-rate-of-change of *F* in an arbitrary control volume V^* (left side) is equal to the inflow of *F* through the boundary plus the source term (right side). The term **b** – **u** is the relative velocity between the control surface and the fluid. When **b** = **u**, we recover the Lagrangian form [Equation (2-1)]. For a control volume that is fixed with respect to the coordinate axes, **b** = 0 and we recover the Eulerian form

$$\frac{\partial}{\partial t}\int_{V^*}\Phi dV=-\oint_{S^*}\Phi(\mathbf{u}\cdot\mathbf{A})dS+\int_{V^*}S_\Phi dV\,, \tag{2-5}$$

or

$$\int_{V^*}\frac{\partial\Phi}{\partial t}dV=-\oint_{S^*}\Phi(\mathbf{u}\cdot\mathbf{A})dS+\int_{V^*}S_\Phi dV\,, \tag{2-6}$$

2.2 The Mass Conservation Equations

The mixture mass conservation equation follows directly from Equation (2-4) by letting $\boldsymbol{\Phi}=\rho$

$$\frac{\partial}{\partial t}\int_{V}\rho dV=\oint_{S}\rho(\mathbf{b}-\mathbf{u})\cdot\mathbf{A}dS+\int_{V}S_\rho dV\,, \tag{2-7}$$

where ρ is the mixture density or the sum of the macroscopic densities for each individual species; **u** is the mass-average velocity vector; and S_ρ is the mass source or sink due to condensation, vaporization, and liquid droplet depletion (rainout) per unit volume and time. Similarly, setting $\boldsymbol{\Phi}$ = 1 in Equation (2-4) gives an expression for the change in volume:

$$\frac{\partial V}{\partial t}=\oint_{S}\mathbf{b}\cdot\mathbf{A}dS\,, \tag{2-8}$$

The transport equation for individual species is given by

$$\frac{\partial}{\partial t}\int_{V}\rho_\alpha dV=\oint_{S}\rho_\alpha(\mathbf{b}-\mathbf{u})\cdot\mathbf{A}dS-\oint_{S}(\mathbf{J}_\alpha\cdot\mathbf{A})dS+\int_{V}S_{\rho,\alpha}dV\,, \tag{2-9}$$

where α denotes the gas species, ρ_α is the mass per unit volume (macroscopic density), $\mathbf{J}_\alpha\cdot\mathbf{A}$ is the mass diffusion flux vector with Cartesian geometry components, $(A_xJ_{x,\alpha}, A_yJ_{y,\alpha}, A_zJ_{z,\alpha})$ and the source or sink term, $S_{\rho,\alpha}$, represents the species mass created or destroyed by chemical reactions and two-phase change dynamics of the liquid and vapor water components.

The diffusion of species α is represented by the second integral on the right side of Equation (2-9), which is often reported in the literature (Ref. 2-4 and Ref. 2-5) as

$$\mathbf{J}_\alpha \cdot \mathbf{A} = \begin{bmatrix} -A_x \rho D_{\alpha\to mix} \dfrac{\partial}{\partial x}\left(\dfrac{\rho_\alpha}{\rho}\right), \\ -A_y \rho D_{\alpha\to mix} \dfrac{\partial}{\partial y}\left(\dfrac{\rho_\alpha}{\rho}\right), \\ -A_z \rho D_{\alpha\to mix} \dfrac{\partial}{\partial z}\left(\dfrac{\rho_\alpha}{\rho}\right) \end{bmatrix} \tag{2-10}$$

for Cartesian geometry, where $D_{\alpha\to mix}$ is the mass diffusion coefficient of species α into the gaseous mixture. The reader is referred to Sections 2.5.1.2 (Transport Properties) and 2.7.4 (Turbulence Effects on the Transport Coefficients) to see what models are used to compute molecular mass diffusion coefficients of the binary coefficients and what effects turbulence has on these coefficients, respectively.

It is essential that the diffusion fluxes identically sum to zero, so Equation (2-10) is

$$\sum_\alpha \mathbf{J}_\alpha \cdot \mathbf{A} = \sum_\alpha \mathbf{J}_\alpha \equiv 0 \tag{2-11}$$

As pointed out by Ramshaw (Ref. 2-6), the condition of Equation (2-11) is only exact for Equation (2-11) when binary mixtures are involved or for multicomponent mixtures when all the diffusion coefficients are equal. This is in general not the case (see Ref. 2-6), so a correction to Equation (2-10) is incorporated that properly reflects the physical content of Equation (2-10) while providing full symmetry in all species:

$$\mathbf{J}_\alpha \cdot \mathbf{A} = \left\{ \begin{matrix} -A_x \left[cM_\alpha D_{\alpha\to mix} \dfrac{\partial}{\partial x}\left(\dfrac{c_\alpha}{c}\right) - \left(\dfrac{c\rho_\alpha}{\rho}\right) \sum_\beta M_\beta D_{\beta\to mix} \dfrac{\partial}{\partial x}\left(\dfrac{c_\beta}{c}\right) \right], \\ -A_y \left[cM_\alpha D_{\alpha\to mix} \dfrac{\partial}{\partial y}\left(\dfrac{c_\alpha}{c}\right) - \left(\dfrac{c\rho_\alpha}{\rho}\right) \sum_\beta M_\beta D_{\beta\to mix} \dfrac{\partial}{\partial y}\left(\dfrac{c_\beta}{c}\right) \right], \\ -A_z \left[cM_\alpha D_{\alpha\to mix} \dfrac{\partial}{\partial z}\left(\dfrac{c_\alpha}{c}\right) - \left(\dfrac{c\rho_\alpha}{\rho}\right) \sum_\beta M_\beta D_{\beta\to mix} \dfrac{\partial}{\partial z}\left(\dfrac{c_\beta}{c}\right) \right] \end{matrix} \right\} \tag{2-12}$$

When Equation (2-9) is summed over all species, the result is the mixture mass equation [Equation (2-7)].

2.3 The Momentum Transport Equations

The mixture-momentum conservation equations are given by

$$\frac{d}{dt}\int_V \rho \mathbf{u} dV = \oint_S \rho \mathbf{u}(\mathbf{b}-\mathbf{u})\cdot \mathbf{A} dS$$

$$-\oint_S p d\mathbf{S} + \int_V \rho \mathbf{g} dV - \oint_S (\tau \cdot \mathbf{A}) dS - \oint_S (\mathbf{D}_d \cdot \mathbf{A}) dS + \int_V \mathbf{S}_m dV, \qquad (2\text{-}13)$$

where p is the pressure, τ is the viscous stress tensor, **g** is the gravitational vector, $\mathbf{D}_d$ is the internal structure drag tensor, and $\mathbf{S}_m$ any momentum sources such as fans. The right-side integrals represent, respectively, the flux of momentum through the control surface; the sum of pressure gradient, gravity, and viscous forces on the control volume; fluid drag forces acting on structural surfaces; and any additional momentum sources.

The Cartesian and cylindrical components of τ for a Newtonian fluid are given in Table 2-1 and Table 2-2, respectively. The components of **u** are (u, v, w) in the x-, y-, and z-directions (Cartesian coordinates) or r-, θ-, and z-directions (cylindrical coordinates); subscripts on τ indicate the surface normal direction and the direction of the stress component.

The reader is referred to Sections 2.5.1.2 (Transport Properties) and 2.7.4 (Turbulence Effects on the Transport Coefficients) to see how the mixture viscosity is computed and what effects turbulence has on the viscosity, respectively. Here we have used the second viscosity coefficient, $\lambda = -2\mu/3$, which is equivalent to assuming the bulk viscosity to be zero.

Table 2-1 Stress tensor components for a Newtonian fluid (Cartesian coordinates)

$\tau_{xx} = -\mu\left[2\frac{\partial u}{\partial x} - \frac{2}{3}(\nabla\cdot\mathbf{u})\right]$	$\tau_{xy} = -\mu\left[\frac{\partial u}{\partial y} + \frac{\partial v}{\partial x}\right]$
$\tau_{yy} = -\mu\left[2\frac{\partial v}{\partial y} - \frac{2}{3}(\nabla\cdot\mathbf{u})\right]$	$\tau_{xz} = -\mu\left[\frac{\partial u}{\partial z} + \frac{\partial w}{\partial x}\right]$
$\tau_{zz} = -\mu\left[2\frac{\partial w}{\partial z} - \frac{2}{3}(\nabla\cdot\mathbf{u})\right]$	$\tau_{yz} = -\mu\left[\frac{\partial v}{\partial z} + \frac{\partial w}{\partial y}\right]$

Table 2-2 Stress tensor components for a Newtonian fluid (cylindrical coordinates)

$\tau_{rr} = -\mu\left[2\frac{\partial u}{\partial r} - \frac{2}{3}(\nabla\cdot\mathbf{u})\right]$	$\tau_{r\theta} = -\mu\left[r\frac{\partial}{\partial r}\left(\frac{v}{r}\right) + \frac{1}{r}\frac{\partial u}{\partial \theta}\right]$
$\tau_{\theta\theta} = -\mu\left[2\left(\frac{1}{r}\frac{\partial v}{\partial \theta} + \frac{u}{r}\right) - \frac{2}{3}(\nabla\cdot\mathbf{u})\right]$	$\tau_{z\theta} = -\mu\left[\frac{\partial v}{\partial z} + \frac{1}{r}\frac{\partial w}{\partial \theta}\right]$
$\tau_{zz} = -\mu\left[2\frac{\partial w}{\partial z} - \frac{2}{3}(\nabla\cdot\mathbf{u})\right]$	$\tau_{rz} = -\mu\left[\frac{\partial u}{\partial z} + \frac{\partial w}{\partial r}\right]$

2.4 The Internal Energy Transport Equation

The equation of change for total internal energy is

$$\frac{d}{dt}\int_V \rho I dV = \oint_S \rho I(\mathbf{b}-\mathbf{u})\cdot \mathbf{A}dS$$

$$-\oint_S p(\mathbf{u}\cdot\mathbf{A})dS - \int_V \left[\frac{p}{V}\frac{\partial V_{h_2o}}{\partial t}\right]dV - \oint_S (\mathbf{q}\cdot\mathbf{A})dS + \int_V S_I dV , \quad (2\text{-}14)$$

where I is the mixture specific internal energy, and $\mathbf{S}_I$ is the energy source or sink per unit volume and time as a result of combustion, phase change, and energy exchange with internal structures, floors, ceilings, and walls. Because phase-change effects can dominate the p_$\mathbf{u}$ work term, we must account for the remaining gas in a computational cell expanding or compressing into the volume change associated with the phase change process. We account for this effect by using the ideal gas equation of state to arrive at

$$\frac{p}{V}\frac{\partial V_{h_2o}}{\partial t} = -R_{h_2o} T S_{\rho,h_2o} ,$$

where R_{h_2o} is the gas constant for steam, T is the gas mixture temperature, V_{h_2o} is the steam volume, and S_{ρ,h_2o} is the sum of all steam mass per unit time lost or gained due to phase change in the fluid mixture and to mass transfer on all surfaces internal to or bounding the computational cell V. Note that S_{ρ,h_2o} is the same mass source per unit volume and time as Equation (2-7), i.e., the water vapor component conservation equation.

The energy flux vector $\mathbf{q}$ is given by

$$\mathbf{q}\cdot\mathbf{A} = \begin{Bmatrix} -A_x\left[\phi_x \dfrac{\partial T}{\partial x} - \sum_\alpha h_\alpha J_{x,\alpha}\right], \\ -A_y\left[\phi_y \dfrac{\partial T}{\partial y} - \sum_\alpha h_\alpha J_{y,\alpha}\right], \\ -A_z\left[\phi_z \dfrac{\partial T}{\partial z} - \sum_\alpha h_\alpha J_{z,\alpha}\right] \end{Bmatrix}, \quad (2\text{-}15)$$

where ϕ_i is the molecular conductivity and h_α is the enthalpy for species α. The reader is referred to Sections 2.5.1.2 (Transport Properties) and 2.7.4 (Turbulence Effects on the Transport Coefficients) to see what models are used to compute the mixture conductivity and what effects turbulence has on these coefficients, respectively.

2.5 Constitutive Relationships

2.5.1 Thermochemical and Transport Properties

GASFLOW-MPI has incorporated a library of 25 species, which are listed in Table 2-3. Two of them are gas mixtures: (1) air, a mixture of 79% nitrogen and 21% oxygen and (2) lg, a "light-gas" consisting of a mixture of 15% hydrogen and 85% helium. Some of the species are not stable molecules, for example, h, o, and oh, but they are included because of the possibility of implementing detailed hydrogen-oxygen chemical kinetics mechanisms which require intermediate reaction species.

Two of the species are not gases: (1) h_2o_2, which is the water liquid component required for the two-phase homogenous equilibrium model and (2) c, which is solid carbon or soot being produced in the event of hydrocarbon combustion.

2.5.1.1 Internal Energy and Specific Heats

The specific internal energy of species α is related to the temperature by

$$I_\alpha = \left(I_{reference}\right)_\alpha + \int_{T_{reference}}^{T} \left(C_v\right)_\alpha dT , \qquad (2\text{-}16)$$

where we have arbitrarily defined our reference temperature 298.15 K and set the reference internal energy equal to zero. (Note that in future versions of GASFLOW-MPI we will not have this limitation, but for the present version, we include the internal energy in this way).

One can approximate the specific heat in Equation (2-16) by polynomials of various degrees. In fact, Gordon and McBride (Ref. 2-8), basing their approximations on the JANAF Tables (Ref. 2-9), give specific heats at constant pressure as a function of temperature in the form of least squares coefficients as follows:

Table 2-3 Gas properties at temperature equaling 298.15 K [extracted from the JANAF (Ref. 2-7) Tables]

Species	Molecular Weight M_α (g/mole)	Gas Constant $\Re_\alpha$ (J/g-K)	Specific Heat $C_{v,\alpha}$ (J/g-K)	Enthalpy of Formation $h^o_{f,\alpha}$ (J/g)
h_2	2.01588	4.12416	10.179479	0.000
o_2	31.99880	0.25983	0.658068	0.000
n_2	28.01340	0.29680	0.742878	0.000
air	28.85033	0.28819	0.723125	0.000
he	4.00260	2.07705	3.115522	0.000
lg	3.70459	2.24411	3.690719	0.000
h_2o	18.01528	0.46150	1.402215	−13422.869
oh	17.00734	0.48886	1.274294	2320.372

h	1.00794	8.24856	12.373200	216262.790
o	15.99940	0.51964	0.849832	15574.679
ho_2	33.00674	0.25189	0.805014	633.786
h_2o_2	34.01468	0.24442	1.023102	–4001.221
nh_3	17.03052	0.48820	1.603728	–2695.058
nh_2	16.02258	0.51891	1.576402	10471.333
nh	15.01464	0.55374	1.387406	22571.525
hno	31.01404	0.26807	0.848757	3210.575
n_2o	44.01280	0.18889	0.688477	1864.055
no	30.00610	0.27707	0.717481	3008.888
ch_4	16.04276	0.51825	1.703236	–4667.000
co	28.01040	0.29683	0.743551	–3946.416
co_2	44.00980	0.18892	0.654727	–8941.658
ar	39.94800	0.20813	0.312192	0.000
xe	131.29	0.06333	0.094993	0.000
h_2ol	18.01528	0.00000	4.179300	–15865.897
c	12.01100	0.00000	0.017708	0.000

$$\frac{C_p^o}{\Re} = a_1 + a_2 T + a_3 T^2 + a_4 T^3 + a_5 T^4 . \tag{2-17}$$

We have based the GASFLOW-MPI approximations for the specific heats at constant volume upon polynomials of up to the third degree. When Equation (2-16) is then integrated for the species' specific internal energy, we approximate the resulting fifth-degree polynomial with a polynomial of up to the fourth degree,

$$I_\alpha = a_\alpha + b_\alpha T + c_\alpha T^2 + d_\alpha T^3 + e_\alpha T^4 , \tag{2-18}$$

which allows analytically inverting this function for the temperature field when the internal energy and species' densities are known. Therefore, GASFLOW-MPI provides the possibility of internal energy as a function of temperature representation from linear to the fourth-degree polynomial shown in Equation (2-18). It must be stated that the selection of accuracy, i.e., the degree of the polynomial, is related to the computer time to invert Equation (2-18), and the user will be reminded of this fact in the GASFLOW-MPI User's Manual (NUREG/CR-6570, Vol. 2).

The total specific internal energy then is given by

$$I = \sum_\alpha x_\alpha I_\alpha = \sum_\alpha x_\alpha a_\alpha + T \sum_\alpha x_\alpha b_\alpha + T^2 \sum_\alpha x_\alpha c_\alpha + T^3 \sum_\alpha x_\alpha d_\alpha + T^4 \sum_\alpha x_\alpha e_\alpha =$$

$$\sum_\alpha x_\alpha a_\alpha + T \left\{ \sum_\alpha x_\alpha b_\alpha + T \left[\sum_\alpha x_\alpha c_\alpha + T \left(\sum_\alpha x_\alpha d_\alpha + T \sum_\alpha x_\alpha e_\alpha \right) \right] \right\} , \tag{2-19}$$

where x_α is the mass fraction for species α. In order that homogeneous equilibrium model thermodynamics be correct, the expression for the liquid water specific internal energy is derived such that the difference between the vapor and liquid specific internal energies is matched by the same difference between the vapor and liquid saturation curve from the steam tables (Ref. 2-10).

We recommend, for consistency between Equations (2-17) and (2-18), that each species' specific heat be obtained by differentiating Equation (2-18) with respect to temperature, but there are other possibilities that will be described in the GASFLOW-MPI User's Manual (NUREG/CR-6570, Vol. 2).

The interrelationships between enthalpy, internal energy, and specific heats are given by the usual thermodynamic relations

$$\begin{aligned} I_\alpha &= h_\alpha - R_\alpha T \\ C_{v,\alpha} &= C_{p,\alpha} - R_\alpha \end{aligned} .$$

2.5.1.2 Transport Properties

The molecular transport properties, i.e., heat conductivities, dynamic viscosities, and binary diffusion coefficients, for all gaseous species are included in the GASFLOW-MPI properties library. We have used the data base of the CHEMKIN (Ref. 2-11) code to provide the GASFLOW-MPI relationships. We use the CHEMKIN model

$$\ln(\Psi) = a_o + a_1 \ln(T) + a_2 \ln(T)^2 + a_3 \ln(T)^3 + a_4 \ln(T)^4 \tag{2-20}$$

for the desired transport property Ψ, as well as perhaps less accurate, but simplified polynomials discussed in the GASFLOW-MPI User's Manual (NUREG/CR-6570, Vol. 2).

When the individual transport properties are determined, the mixture values can also be computed. For the mixture thermal conductivity, the relationship of Mathur et al. (Ref. 2-12), is used:

$$\phi = \frac{1}{2}\left(\sum_{\alpha=1}^{N} Y_\alpha \phi_\alpha + \frac{1}{\sum_{\alpha=1}^{N} Y_\alpha / \phi_\alpha} \right) , \tag{2-21}$$

The modified semi-empirical formulas of Wilke (Ref. 2-13), modified by Bird et al. (Ref. 2-4), can be used to compute the mixture viscosity as

$$\mu = \sum_{\alpha=1}^{N} \frac{Y_\alpha \mu_\alpha}{\sum_{\beta=1}^{N} Y_\alpha \Phi_{\alpha\beta}} , \tag{2-22}$$

where

$$\Phi_{\alpha\beta} = \frac{1}{\sqrt{8}}\left(1+\frac{M_\alpha}{M_\beta}\right)^{-\frac{1}{2}}\left[1+\left(\frac{\mu_\alpha}{\mu_\beta}\right)^{\frac{1}{2}}\left(\frac{M_\beta}{M_\alpha}\right)^{\frac{1}{4}}\right]^2 \quad . \tag{2-23}$$

We determine the diffusion coefficient of species α into the mixture by the following well-known relationship (Ref. 2-14):

$$D_{\alpha\rightarrow mixture} = \frac{\sum\limits_{\beta\neq\alpha} Y_\beta}{\sum\limits_{\beta\neq\alpha} Y_\beta / D_{\alpha\rightarrow\beta}} \quad . \tag{2-24}$$

2.5.2 Equation of State

The pressure field, *p*, is obtained by applying the Gibbs-Dalton law of partial pressures to an ideal gas mixture in the available gas volume (void volume):

$$p = \frac{T\sum\limits_{\alpha}^{\alpha\neq h_2ol} R_\alpha \rho_\alpha}{\theta} = \frac{\rho T \sum\limits_{\alpha}^{\alpha\neq h_2ol} R_\alpha x_\alpha}{\theta} \quad , \tag{2-25}$$

where R_α is the gas constant for species α. The volume fraction, θ, of the liquid water phase is computed from

$$\theta = 1 - \frac{\rho_{h_2ol}}{\rho_{h_2ol}^{thermo}} \quad , \tag{2-26}$$

where the thermodynamic liquid water density, $\rho_{h_2ol}^{thermo}$, is usually approximated as a constant value of 1 g/cm^3.

2.6 Heat-Transfer and Phase-Change Relationships

2.6.1 Gas-Structure Heat Transfer

The convective heat exchange between the gas mixture and a solid boundary (referred to generically as a wall, but it may in fact be, in addition, a ceiling, floor, or internal structure) is given by

$$\int\limits_V S_{I,convection} dV = \sum\limits_s h_s A_s (T_s - T), \tag{2-27}$$

where T_s is the structure surface temperature, *T* is the gas temperature, h_s is the heat-transfer coefficient between the gas mixture and the internal structures, and A_s is the cell face area for walls or the exposed area for internal structures in a computational cell. The thermal energy delivered to the wall surface, $q_{s,convection}$, has the opposite sign of the gas mixture, so the thermal energy becomes

$$q_{s,convection} = h_s A_s \left(T - T_s\right). \tag{2-28}$$

The thermal boundary layer is taken into account by using a modified Reynolds analogy formulation (Ref. 2-15), which is simplified and combined with a Chilton-Colburn empirical analogy (Ref. 2-4) between the momentum and thermal boundary layers to obtain the heat-transfer coefficient:

$$h_s = \frac{\tau_s}{|\mathbf{u}_c|} C_p \cdot \Pr^{-\frac{2}{3}}. \tag{2-29}$$

See Section 2.6.6 (Wall Shear Stress) below for a more detailed discussion of this heat-transfer coefficient.

The rates of heat transfer and condensation increase when the mass fraction of steam becomes a relatively large fraction of the mass of the gas mixture. As the mass-transfer rate increases, the thermal and concentration boundary layers become thinner because of the suction effect of the condensation process. This reduction in the boundary layer thickness further increases the temperature and concentration gradients near the boundary and consequently increases the heat- and mass-transfer coefficients. The opposite effect occurs at a surface where vaporization of a liquid film is taking place; hence, the heat- and mass-transfer coefficients decrease during these conditions. Bird, Stewart, and Lightfoot (Ref. 2-4, Section 2.5.1) develop correction factors based on film theory that can be used to determine the increase in the heat- and mass-transfer coefficients. The corrected heat-transfer coefficient then becomes

$$h_s^* = \Theta_T h_s \quad , \tag{2-30}$$

where

$$\Theta_T = \frac{\phi_T}{e^{\phi_T} - 1} \tag{2-31}$$

and the rate factor, ϕ_T, is given by

$$\phi_T = \frac{-\dot{m}_s C_{p,h_2o}}{h_s A_s} \quad , \tag{2-32}$$

where $\dot{m}_s$ is the wall condensation or vaporization rate, equal to one of the surface contributions involving the term $\int_V S_{p,h_2o} dV$ in Equation (2-9), and C_{p,h_2o} is the specific heat of the water vapor at constant pressure. Note that in the presence of condensing water vapor, ϕ_T is negative which increases the correction factor, Θ_T, and the heat-transfer coefficient, h_s^*; the opposite effect occurs during evaporation of a surface film when ϕ_T is positive. The internal structure heat-transfer coefficient is computed in an analogous fashion.

2.6.2 Condensation and Vaporization

Phase change between the gas and structures can occur on any structural surface (walls, ceiling, floors, and internal structures) can occur under one of these two conditions: (1) the surface temperature is less than the saturation temperature of the water vapor next to the surface (condensation), or (2) condensate exists on any given structural surface and the condensate surface (i.e., the surface temperature as well) is greater than the saturation temperature of the gas mixture adjacent to the surface (vaporization).

The phase-change rate on any structural surface is described as

$$\dot{m}_s = h_d^* A_s \left(\rho_{h_2o} - \rho_{s,saturation} \right) \quad , \tag{2-33}$$

where h_d^* is the corrected mass-transfer coefficient, ρ_{h_2o} is the water vapor density in the gas mixture, and $\rho_{s,saturation}$ is the saturation water vapor density at the structural surface conditions. h_d^* is the corrected mass-transfer coefficient. ρ_{h_2o} is the water vapor density in the gas mixture.

The saturation density in the above relationship is computed from the saturation pressure and the structural surface temperature by

$$\rho_{s,saturation}\left[T_s, p_{s,saturationt}\left(T_s\right)\right] = \frac{p_{s,saturation}(T_s)}{R_{h_2o} \cdot T_s} \quad , \tag{2-34}$$

where the saturation pressure as a function of temperature is evaluated from the integrated Clausius-Clapeyron equation for the water component saturation curve in the form

$$p_{sat}(T) = 10^6 \cdot e^{\frac{-c_1 + c_2 \cdot T}{c_3 \cdot T}} \quad , \tag{2-35}$$

and the coefficients are defined

$$\begin{aligned} c_1 &= 2258.0 \\ c_2 &= 6.05963 \\ c_3 &= 0.4579742 \end{aligned} \quad . \tag{2-36}$$

The mass-transfer coefficient, h_d, then can be expressed in terms of the heat-transfer coefficient, h_S (Ref. 2-16), as

$$h_d = \frac{h_s}{\rho C_p} \frac{Sc^{-\frac{2}{3}}}{\Pr^{-\frac{2}{3}}} \tag{2-37}$$

when we make use of a Chilton-Colburn empirical analogy between heat and mass transfer (Ref. 2-4).

Following similar ideas as with the heat-transfer coefficient for relatively large steam mass fractions, we correct the mass-transfer coefficient by

$$h_d^* = \Theta_m h_d \quad , \tag{2-38}$$

where

$$\Theta_m = \frac{\log(R+1)}{R} \tag{2-39}$$

and the flex ratio R is expressed as

$$R = \frac{n_{s,h_2o} - n_{h_2o}}{1 - n_{s,h_2o}} \tag{2-40}$$

where n_{s,h_2o} is the steam mole fraction at the wall, and n_{h_2o} is the steam mole fraction in the gas mixture.

For the situation where "dryout" of a surface may occur, i.e., the liquid film totally evaporates leaving the surface dry, a better formulation of the surface mass transfer equation (2-33) is

$$\dot{m}_s = \max\left[-\frac{\delta A_s \rho_{h_2ol}^{thermo}}{2\Delta t}, h_d^* A_s \left(\rho_{h_2o} - \rho_{s,saturation}\right) \right] \quad , \tag{2-41}$$

where δ is the film thickness and Δt is the time increment of the computational time step in seconds. Note the sign of Equation (2-41): positive indicates condensation, whereas negative means vaporization of the liquid film, where, at the point of near dryout, we allow only half of the available film to evaporate in a time step.

The total mass source or sink term due to phase change involving structural surfaces in Equations (2-7) and (2-9) for all surfaces then becomes

$$\int_V S_{\rho,h_2o,condensation/vaporization} dV = -\sum_s \max\left[-\frac{\delta A_s \rho_{h_2ol}^{thermo}}{2\Delta t}, h_d^* A_s \left(\rho_{h_2o} - \rho_{s,saturation}\right) \right] \quad , \tag{2-42}$$

The amount of energy resulting from phase change and the structural wall surface is then

$$q_{s,condensation/vaporization} = \max\begin{bmatrix} -\frac{\delta A_s \rho_{h_2ol}^{thermo}}{2\Delta t} I_{h_2o}(T_s), \\ h_d^* A_s \left(\rho_{h_2o} - \rho_{s,saturation}\right) I_{h_2o}(T_s), \\ h_d^* A_s \left(\rho_{h_2o} - \rho_{s,saturation}\right) I_{h_2o}(T) \end{bmatrix} \quad , \tag{2-43}$$

where $I_{h_2o}(T)$ is the specific internal energy of the water vapor in the computational cell adjacent to the wall with volume V_c, and $I_{h_2o}(T_s)$ is the specific internal energy of the liquid water film that is on the surface. (Note that we assume the film temperature is equal to the surface temperature of the wall.) In this case, the energy lost or gained in the gas mixture is

$$\int_V S_{I,condensation/vaporization} dV = -\sum_s \max \begin{bmatrix} -\frac{\delta A_s \rho_{h_2ol}^{thermo}}{2\Delta t} I_{h_2o}(T_s), \\ h_d^* A_s \left(\rho_{h_2o} - \rho_{w,saturation}\right) I_{h_2o}(T_s), \\ h_d^* A_s \left(\rho_{h_2o} - \rho_{s,saturation}\right) I_{h_2o}(T) \end{bmatrix} . \quad (2\text{-}44)$$

2.6.3 Phase Change in the Fluid Mixture – The Homogeneous Equilibrium Model (HEM)

The nonequilibrium phase exchange function between the vapor and liquid phases is presented here as a relaxation type function. The mass exchange between the phases, shown as a source and sink term in Equation (2-9), is given for the water vapor phase by

$$\int_V S_{\rho,h_2o,fluid} dV = VC\left[\rho_{saturation}(T, p_{saturation}) - \rho_{h_2o}\right] \quad (2\text{-}45)$$

and for the water liquid phase as

$$\int_V S_{\rho,h_2ol,fluid} dV = -VC\left[\rho_{saturation}(T, p_{saturation}) - \rho_{h_2o}\right] , \quad (2\text{-}46)$$

where the saturation pressure is again given by

$$p_{saturation}(T) = 10^6 \cdot e^{\frac{-c_1 + c_2 \cdot T}{c_3 \cdot T}} \quad (2\text{-}47)$$

and the saturation density is then computed by

$$\rho_{saturation}(T, p_{saturation}) = \frac{p_{saturation}(T)}{R_{h_2o} \cdot T} . \quad (2\text{-}48)$$

The relaxation coefficient C, which has units of inverse time, is a user input value [see **cbulkrlx** in the GASFLOW-MPI User's Manual (NUREG/CR-6570, Vol. 2)], but GASFLOW-MPI checks to see if this input value produces an acceptable stable solution for the current time step.

The work due to the production or loss of steam on the internal energy control volume in Equation (2-12) can now be computed from

$$-\int_V \left[\frac{p}{V} \frac{\partial V_{h_2o}}{\partial t}\right] dV = \int_V \left[R_{h_2o} T S_{h_2o}\right] dV = R_{h_2o} T \int_V S_{h_2o} dV =$$

$$R_{h_2o}T\left[\int_V S_{\rho,h_2o,condensation/vaporization}dV+\int_V S_{\rho,h_2o,fluid}dV\right]=$$

$$R_{h_2o}T\left\{\begin{array}{l}-\sum_s \max\left[-\frac{\delta A_s\rho_{h_2ol}^{thermo}}{2\Delta t},h_d^* A_s\left(\rho_{h_2o}-\rho_{s,saturation}\right)\right]\\+VC\left[\rho_{saturation}(T,p_{saturation})-\rho_{h_2o}\right]\end{array}\right\}\quad . \tag{2-49}$$

2.6.4 Droplet Depletion or Droplet "Rainout"

When the nonequilibrium phase exchange function between the vapor and liquid phases is active, conditions may exist that could produce large amounts of liquid water (water droplets). In these cases, we would like to have the option of depleting the liquid water mass in the fluid mixture when a certain droplet loading has occurred. This can be done with a relaxation type function, where the liquid droplet mass, when the liquid water exceeds a certain loading, relaxes toward the accepted loading value. The loss or sink term to account for this event in Equation (2-9) is given for the liquid phase by

$$\int_V S_{\rho,h_2ol,mixture}dV=VC_{h_2ol,mixture}\cdot\min\left[0,\left(\rho_{h_2ol,\max}-\rho_{h_2ol}\right)\right]\quad . \tag{2-50}$$

The relaxation coefficient, $C_{h_2ol,mixture}$, which has units of inverse time, and the maximum loading density, $\rho_{h_2ol,\max}$, are user input values [see **crelax** and **rholiqmx,** respectively, in the GASFLOW-MPI User's Manual (NUREG/CR-6570, Vol. 2)], but GASFLOW-MPI checks to see if these input values produce an acceptable stable solution for the current time step.

The loss of droplets on the internal energy control volume in Equation (2-12) can now be computed from

$$\int_V S_{I,rainout}dV=VC_{h_2ol,mixture}\cdot\min\left[0,\left(\rho_{h_2ol,\max}-\rho_{h_2ol}\right)\right]\cdot I_{h_2ol}(T)\ , \tag{2-51}$$

2.6.5 Structural Heat Conduction

For every computational cell side interfacing with a wall, ceiling, or floor and any defined distributed heat sinks, the one-dimensional transient heat-conduction equation

$$\frac{\partial}{\partial t}\int_V q\,dV=\int_S k\frac{\partial T}{\partial x}dS \tag{2-52}$$

with the wall boundary condition from Equations (2-28) and (2-43)

$$q_{s,convection} + q_{s,condensation/vaporization} = -kA_s \left.\frac{\partial T_w}{\partial x}\right|_{x=s} \tag{2-53}$$

is solved from the temperature distribution, $T_W(\mathbf{x},t)$, and the wall surface temperature, T_S. The term k in Equations (2-52) and (2-53) is the thermal conductivity of the structure. On the left side of Equation (2-53) the two terms represent energy delivered to a wall section by convection and phase change, respectively.

2.6.6 Wall Shear Stress

The heat-transfer coefficient expression [Equation (2-29)] contains the computational cell-centered average velocity, $\mathbf{u}_c$, a vector with the two wall tangential velocity components, and the wall shear stress, τ_s, which is related to the fluid density and the wall shear speed, u_*, by

$$\tau_s = \rho u_*^2 \quad . \tag{2-54}$$

We are unable to resolve turbulent boundary layers near solid walls with any practical computing mesh, so we match our solution near solid boundaries or internal structures with the turbulent law-of-the-wall (Ref. 2-17):

$$\frac{|\mathbf{u}_c|}{u_*} = A \ln\left(\frac{y_c u_*}{\nu}\right) + B \quad . \tag{2-55}$$

This expression requires an iterative solution for u_*. We find that it is more convenient and almost as accurate to use an approximation obtained by replacing u_* in the argument of the logarithm in Equation (2-55) by the one-seventh-power law (Ref. 2-18). The one-seventh-power law may be rearranged to give

$$\frac{y_c u_*}{\nu} = 0.15\left(\frac{y_c|\mathbf{u}_c|}{\nu}\right)^{\frac{7}{8}} \quad , \tag{2-56}$$

which yields

$$\frac{|\mathbf{u}_c|}{u_*} = 2.19 \ln\left(\frac{y_c|\mathbf{u}_c|}{\nu}\right) + 0.76 \tag{2-57}$$

when substituted into Equation (2-55) and when A=2.5 and B=5.5. It is now straightforward to find the shear speed, u_*, where y_c is the distance from the wall to the cell-centered average tangential speed, $|\mathbf{u}_c|$, and ν is the gas mixture molecular kinematic viscosity.

The local Reynolds number, ($y_c |\mathbf{u}_c|/\nu$), may be small, indicating that the cell center lies in the laminar sub layer and the law-of-the-wall formulation is not valid. In this case, Equation (2-57) is replaced by the corresponding laminar formula:

$$\frac{|\mathbf{u}_c|}{u_*} = \left(\frac{y_c |\mathbf{u}_c|}{\nu} \right)^{\frac{1}{2}} \quad . \tag{2-58}$$

The transition between Equations (2-57) and (2-58) is made at the value where they predict the same wall surface shear speed, u_*, which is $(y_c |\mathbf{u}_c|/\nu)$ = 130.7. Therefore, u_* is calculated by Equation (2-57) when $(y_c |\mathbf{u}_c|/\nu)$ _ 130.7 and by Equation (2-58) when $(y_c |\mathbf{u}_c|/\nu) < 130.7$. In the laminar case, the wall heat-transfer coefficient [Equation (2-29)] reduces to $h_w = \rho\nu\, c_p / y_c$, which results in a simple difference approximation to the laminar heat flux for a molecular Prandtl number of unity when substituted into Equation (2-28).

Therefore, the uncorrected heat-transfer coefficient calculated from Equation (2-29) becomes

$$h_s = \begin{cases} \dfrac{\nu\rho C_p}{y_c}, & \dfrac{y_c |\mathbf{u}_c|}{\nu} < 130.7 \\ \dfrac{\rho C_p |\mathbf{u}_c|}{\left[2.19 \ln\left(\dfrac{y_c |\mathbf{u}_c|}{\nu} \right) + 0.76 \right]^2} \mathrm{Pr}^{-\frac{2}{3}}, & \dfrac{y_c |\mathbf{u}_c|}{\nu} \geq 130.7 \end{cases} \quad . \tag{2-59}$$

2.7 Turbulence Modeling

Most flows of engineering interest, including the flows of concern here, are turbulent. Turbulence may be described qualitatively as the superposition of an irregular fluctuating motion on the mean flow, which, for an arbitrary variable *y*, may be expressed as

$$\psi(\mathbf{x},t) = \overline{\psi}(\mathbf{x}) + \psi'(\mathbf{x},t) \quad , \tag{2-60}$$

where ψ is the instantaneous value, ψ' is the fluctuating (turbulent) component, and $\overline{\psi}$ is the mean value defined as time average

$$\overline{\psi}(\mathbf{x}) = \lim_{t\to\infty} \frac{1}{t} \int_{t_0}^{t_0+t} \psi(\mathbf{x},t)dt \quad . \tag{2-61}$$

In practice, *t* is taken to be much larger than the characteristic times associated with ψ'. Formally, this time Reynolds averaging is only appropriate for stationary turbulence (Ref. 2-19). In practice, virtually all the problems involving turbulent flows that we are interested in are classified as inhomogenous turbulent flows. Therefore, in order to make use of the Reynolds-averaged equations, we assume nearly stationary or quasi-stationary turbulent conditions, which are discussed below.

When the instantaneous values of *p, ρ, μ, and* u_i ($u_i \equiv u, v, w$ for i = 1, 2, 3) are substituted into the mixture-momentum equations [Equation (2-13)], the resulting equations contain additional products of terms involving u_i', $\bar{u}_i$, ρ', and $\bar{\rho}$. The terms $-\rho\overline{u_i' u_j'}$, called the Reynolds stresses, are the focus of attention in the turbulent equations. [Other correlations such as $\overline{\bar{u}_i \rho' u_j'}$ are discussed, for example, by Cebeci and Smith (Ref. 2-20, Chapter 0).] Defining $-\bar{\rho}\overline{u_i'u_j'}$ as the components of the turbulent stress tensor τ_t and combining them with the laminar viscous stress, $\bar{\tau}$, the total stress tensor may be written as

$$\boldsymbol{\tau} = \bar{\boldsymbol{\tau}} + \boldsymbol{\tau}_t \ . \tag{2-62}$$

The effect of τ_t in most flows is a large increase in the apparent (turbulent) resistance of the flow; in other words, τ_t results in increased momentum transport. With the addition of the unknown turbulence quantities, Equations (2-9), (2-13) and (2-14) no longer form a closed set and modeling τ_t becomes the major problem in simulating turbulent motion.

To understand the general approach to modeling τ_t, it is useful to consider two observations about the eddies that characterize the turbulent motion. First, the largest eddies (whose size is determined by the geometry of the flow) carry most of the turbulent kinetic energy. The smallest eddies, with sizes determined by molecular viscosity, dissipate turbulent kinetic energy. The qualitative dynamics of the eddies and their interaction with the mean flow are described by Rodi (Ref. 2-21).

The large eddies interact with mean flow (because the scales of both are similar), thereby extracting kinetic energy from the mean motion and feeding it into the large-scale turbulent motion. The eddies can be considered as vortex elements which stretch each other. Due to this vortex stretching, which is an essential feature of the turbulent motion, the energy is passed on to smaller and smaller eddies until viscous forces become active and dissipate the energy. This process is called energy cascade. The rate at which mean-flow energy is fed into the turbulent motion is determined by the large-scale motion; only this amount of energy can be passed on to smaller scales and finally be dissipated. Therefore, the rate of energy dissipated is also determined by the large-scale motion although dissipation is a viscous process and takes place at the smallest eddies. It is important to note that viscosity does not determine the amount of dissipated energy but only the scale at which dissipation takes place. The smaller the effective viscosity (i.e., the larger the Reynolds number), the smaller are the dissipative eddies relative to the large-scale eddies. When buoyancy forces are present, there is also an exchange between potential energy of the mean flow and turbulent kinetic energy, which can go in both directions but is also effected through the large-scale motion.

The preceding observations are the basis for modeling the effects of turbulence on the mean flow, specifically in choosing the relevant velocity and length scales that characterize the local state of turbulence. Postulating an analogy between laminar stresses and Reynolds stresses (Boussinesq's idea), that is,

$$\left(\tau_{ij}\right)_t = -\bar{\rho}\overline{u_i'u_j'} = \mu_t \frac{\partial u_i}{\partial x_j} \ , \tag{2-63}$$

the turbulence closure problem is seen as one of finding a turbulent or eddy viscosity μ_t. The effects of turbulence on the mean flow come down to modeling μ_t as a function of fluid properties, the dynamics of the flow, and the geometry. The two turbulence models used in GASFLOW-MPI, algebraic and κ-ε, are, respectively, zero- and two-transport-equation models that predict the velocity and length scales that are used to compute μ_t. The models are described in order of increasing complexity, number of equations, and computational effort and, as a rule, in order of accuracy.

Turbulence effects in the vicinity of a wall are modeled with the law-of-the-wall formulation (Section 2.6.5).

2.7.1 Algebraic Model

For a turbulent velocity scale, Prandtl suggested that the square root of the mean turbulent kinetic energy κ is a natural choice, where

$$\kappa = \frac{1}{2}\left(\overline{u'^2} + \overline{v'^2} + \overline{w'^2}\right)^{1/2} \tag{2-64}$$

and a length scale l characteristic of the size of the energy-carrying eddies. Thus,

$$\mu_t = C_\mu \rho \kappa^{1/2} l \quad , \tag{2-65}$$

where C_μ is a constant (typically 0.05). It is often estimated that 10% or less of the mean flow energy is contained in the turbulent kinetic energy, so

$$\kappa^{1/2} = \left[0.1(1/2)u^2\right]^{1/2} \quad . \tag{2-66}$$

For containment studies, the length scale usually is set equal to 0.25~0.5 m.

2.7.2 κ-ε Model

The Navier-Stokes equations may be manipulated to produce exact expressions for κ, the turbulent kinetic energy, and ε, the rate of dissipation of turbulent kinetic energy (Ref. 2-22) defined as

$$\varepsilon = \nu \sum_i \sum_j \left(\overline{\frac{\partial u_i}{\partial x_j}}\right)^2 \quad . \tag{2-67}$$

The exact κ and ε equations are modeled by a pair of approximate transport equations developed by Launder and Spalding (Ref. 2-15), with an extension to treat buoyancy effects. Again, $\kappa^{1/2}$ is the characteristic velocity scale, and the length scale is proportional to $\kappa^{3/2}/\varepsilon$. The transport equation for the product $\rho\kappa$ is given by

$$\frac{\partial}{\partial t}\int_V \rho\kappa dV = \oint_S \left[\rho\kappa(\mathbf{b}-\mathbf{u}) + \left(\frac{\mu}{\sigma_\kappa}\nabla\kappa\right) + \tau : \nabla\mathbf{u}\right] \cdot \mathbf{A} dS$$

$$+ \int_V \left[\mu\alpha\mathbf{g} \cdot \nabla T - \rho\varepsilon + \kappa S_\kappa\right] dV \quad , \tag{2-68}$$

where α is the coefficient of thermal expansion. The advection, diffusion, and shear production of turbulent kinetic energy are given by the three terms in the first integral on the right side; the three terms in the second integral represent its production by buoyancy, dissipation, and generation from sources, respectively.

The transport equation for the product of the density and the dissipation of the turbulent kinetic energy, $\rho\varepsilon$, is

$$\frac{\partial}{\partial t}\int_V \rho\varepsilon dV = \oint_S \left[\rho\varepsilon(\mathbf{b}-\mathbf{u}) + \left(\frac{\mu}{\sigma_\varepsilon}\nabla\varepsilon\right) + C_1\frac{\varepsilon}{\kappa}\tau:\nabla\mathbf{u}\right]\cdot\mathbf{A}dS$$

$$+\int_V \left[C_1\frac{\varepsilon}{\kappa}\mu\alpha\mathbf{g}\cdot\nabla T - C_1\rho\frac{\varepsilon^2}{\kappa} + \varepsilon S_\varepsilon\right]dV . \qquad (2\text{-}69)$$

The terms on the right-hand side have meanings analogous to those in Equation (2-68) The turbulent viscosity is calculated using the Kolmogorov hypothesis:

$$\mu_t = \frac{C_\mu \rho\kappa^{1/2}}{\varepsilon} \quad . \qquad (2\text{-}70)$$

The values of the five new constants (C_1, C_2, C_μ, S_κ, S_ε) appearing in Equations (2-68) and (2-69) and listed in

Table 2-4 Constants used in the κ-ε turbulence model

C_1	C_2	C_μ	σ_κ	σ_ε
1.44	1.92	0.09	1.0	1.3

are those suggested by Launder and Spalding (Ref. 2-15) following an extensive examination of experimental data for free turbulent flows.

2.7.3 SST *κ-ω* Model

warning: The GASFLOW-MPI SST *κ-ω* model is under development as time permits.

The shear stress transport (SST) *κ-ω* model was originally developed by Menter (Ref. 2-50, Ref. 2-51) for accurate prediction of aeronautics flows with strong adverse pressure gradients and separation. It is known that the standard *κ-ω* model shows a strong sensitivity to the initial values of *ω* in free-streams outside of the boundary layer which can be avoided by the *κ-ε* model. This motivation of the SST *κ-ω* model is to ensure a proper selection of *κ-ε* or *κ-ω* zones without user interaction using the blending functions. The main additional complexity is to calculate the distance to the nearest wall which is required in the blending functions.

The standard κ and ω transport equations are:

$$\begin{cases} \dfrac{\partial(\rho\kappa)}{\partial t}+\nabla\bullet(\rho\kappa\mathbf{U})=\nabla\bullet\left[(\mu_l+\sigma_{k1}\mu_t)\nabla\kappa\right]+P_\kappa+P_{\kappa b}-\beta'\rho\kappa\omega \\ \dfrac{\partial(\rho\omega)}{\partial t}+\nabla\bullet(\rho\omega\mathbf{U})=\nabla\bullet\left[(\mu_l+\sigma_{\omega1}\mu_t)\nabla\omega\right]+\dfrac{\omega}{\kappa}\alpha_1(P_\kappa+P_{\kappa b})-\beta_1\rho\omega^2 \end{cases} \tag{2-71}$$

The closure coefficients in Standard κ-ω model are

$$\sigma_{k1}=0.5,\ \beta'=0.09,\ \sigma_{\omega1}=0.5,\ \alpha_1=5/9,\ \beta_1=0.075$$

The standard κ-ε model is converted into a κ-ω formulation:

$$\begin{cases} \dfrac{\partial(\rho\kappa)}{\partial t}+\nabla\bullet(\rho\kappa\mathbf{U})=\nabla\bullet\left[(\mu_l+\sigma_{k2}\mu_t)\nabla\kappa\right]+P_\kappa+P_{\kappa b}-\beta'\rho\kappa\omega \\ \dfrac{\partial(\rho\omega)}{\partial t}+\nabla\bullet(\rho\omega\mathbf{U})=\nabla\bullet\left[(\mu_l+\sigma_{\omega2}\mu_t)\nabla\omega\right]+2\rho\sigma_{\omega2}\dfrac{1}{\omega}\nabla\kappa\bullet\nabla\omega+\dfrac{\omega}{\kappa}\alpha_2(P_\kappa+P_{\kappa b})-\beta_2\rho\omega^2 \end{cases} \tag{2-72}$$

The closure coefficients in Transformed κ-ε model are

$$\sigma_{k2}=1,\ \beta'=0.09,\ \sigma_{\omega2}=0.856,\ \alpha_2=0.44,\ \beta_2=0.0828$$

P_κ is the rate of production of turbulent kinetic energy by the mean flow, namely a transfer of kinetic energy from the mean flow to the turbulence,

$$P_\kappa=-\frac{2}{3}\rho k\nabla\bullet\mathbf{U}-\frac{2\mu_t}{3}(\nabla\bullet\mathbf{U})^2+\mu_t\nabla\mathbf{U}\bullet\left(\nabla\mathbf{U}+(\nabla\mathbf{U})^T\right) \tag{2-73}$$

where μ_t is the turbulent dynamic viscosity. The turbulence production term due to the buoyancy, $P_{\kappa b}$, is

$$P_{\kappa b}=-\frac{\mu_t}{\rho\sigma_b}\mathbf{g}\bullet\nabla\rho \tag{2-74}$$

The turbulent Schmidt number σ_b is 0.7 for full buoyancy model in GASFLOW-MPI.

The equations of standard κ-ω model are multiplied by blending function F_1, the transformed κ-ε equations by a function 1-F_1 and the corresponding κ- and ω- equations are added to read:

$$\begin{cases} \dfrac{\partial(\rho\kappa)}{\partial t}+\nabla\bullet(\rho\kappa\mathbf{U})=\nabla\bullet\left[(\mu_l+\sigma_{k3}\mu_t)\nabla\kappa\right]+P_\kappa+P_{\kappa b}-\beta'\rho\kappa\omega \\ \dfrac{\partial(\rho\omega)}{\partial t}+\nabla\bullet(\rho\omega\mathbf{U})=\nabla\bullet\left[(\mu_l+\sigma_{\omega3}\mu_t)\nabla\omega\right]+(1-F_1)2\rho\sigma_{\omega2}\dfrac{1}{\omega}\nabla\kappa\bullet\nabla\omega+\dfrac{\omega}{\kappa}\alpha_3(P_\kappa+P_{\kappa b})-\beta_3\rho\omega^2 \end{cases} \tag{2-75}$$

The closure coefficients are calculated by

$$\begin{aligned}
\sigma_{k3} &= F_1\sigma_{k1} + (1-F_1)\sigma_{k2}, \\
\sigma_{\omega3} &= F_1\sigma_{\omega1} + (1-F_1)\sigma_{\omega2}, \\
\alpha_3 &= F_1\alpha_1 + (1-F_1)\alpha_2, \\
\beta_3 &= F_1\beta_1 + (1-F_1)\beta_2
\end{aligned}$$

To avoid the over prediction of the eddy viscosity and obtain proper transport behaviour, a limiter is added to the formulation of the eddy-viscosity to account for the transport of the turbulent shear stress,

$$\nu_t = \frac{a_1\kappa}{\max(a_1\omega, SF_2)} \tag{2-76}$$

F_2 is a blending function which restricts the limiter to the wall boundary layer, as the underlying assumptions are not correct for free shear flow. a_1 is a constant equalling 0.31. S is the modulus of the mean strain rate tensor.

The blending functions, F_1 and F_2, which are based on the flow variables and on the distance to the nearest wall are critical to the success of the method. F_1 is defined

$$F_1 = \tanh\left(arg_1^4\right), \tag{2-77}$$

with

$$arg_1 = \min\left(\max\left(\frac{\sqrt{\kappa}}{\beta'\omega y}, \frac{500\nu}{y^2\omega}\right), \frac{4\rho\kappa\sigma_{\omega2}}{CD_{\kappa\omega}y^2}\right), \tag{2-78}$$

where y is the distance to the nearest wall and

$$CD_{\kappa\omega} = \max\left(2\rho\frac{\sigma_{\omega2}}{\omega}\nabla\kappa\bullet\nabla\omega, 10^{-10}\right), \tag{2-79}$$

F_2 is defined

$$F_2 = \tanh\left(arg_2^2\right) \tag{2-80}$$

with

$$arg_2 = \max\left(\frac{2\sqrt{\kappa}}{\beta'\omega y}, \frac{500\nu}{y^2\omega}\right) \tag{2-81}$$

GASFLOW-MPI provides the locations of all wall surfaces. For a specific gas cell, the distance from the cell to all of the no-slip wall surfaces is calculated, and then search the minimum distance. For the flows without any wall effect (without any wall or all the walls are free-slip), in GASFLOW-MPI SST κ-

ω model we assume the distance to the nearest wall is infinite. Then the blending functions $F_1 = F_2 = 0$ which leads to the κ-ε model in κ-ω formulation.

2.7.4 Turbulence Effects on the Transport Coefficients

By postulating the Boussinesq analogy (Ref. 2-18) between molecular stresses and Reynolds stresses, we can replace the molecular diffusion coefficients for mass [Equations (2-9) and (2-12)], energy [Equations (2-14) and (2-16)], and momentum [Equation (2-13), Table 2-1 and Table 2-2] with the sum of the molecular and turbulent values. The resulting diffusion coefficient is often referred to as the "total" or "apparent" diffusivity, thermal conductivity, and viscosity, respectively, for the mass, energy, and momentum diffusion terms. Therefore, we represent these three transport coefficients in the following way:

$$\begin{aligned} D_{\alpha\rightarrow mix} &\rightarrow D_{apparent} = D_{\alpha\rightarrow mix} + D_t \\ \phi &\rightarrow \phi_{apparent} = \phi + \phi_t \\ \mu &\rightarrow \mu_{apparent} = \mu + \mu_t \end{aligned} \tag{2-82}$$

The turbulent conductivity, ϕ_t, is

$$\phi_t = \frac{\mu_t C_p}{\Pr_t} \tag{2-83}$$

and the turbulent mass diffusivity, D_t, is

$$D_t = \frac{\mu_t}{\rho Sc_t} \tag{2-84}$$

which involve the turbulent Prandtl and Schmidt numbers.

The turbulent Prandtl number, Pr_t, can usually be assumed constant. Values of 0.90 for shear flows and 0.50 for free shear layers are usually satisfactory, whereas the turbulent Schmidt number, Sc_t, is normally selected between 0.5 and 1.0. Both the turbulent Prandtl and Schmidt numbers are properties of the flow field and hence are not material properties.

2.8 Chemical Kinetics

2.8.1 One-Step Global Chemical Kinetics Model

A simple one-step global chemical kinetics model that grossly over simplifies the actual chemical processes has been used. In the present implementation of this model, the only reaction modeled is

$$2h_2 + o_2 \xrightarrow{\dot{\omega}} 2h_2o \quad . \tag{2-85}$$

In modeling nuclear reactor containment buildings, typical computational cell volumes are 1–2 m^3; they are larger in some cases. We try to keep cell volumes to about 1 m^3 in regions where diffusion flames are expected. For this spatial resolution, there is no attempt to describe flame structure; we simply represent combustion energy release in a complex geometric containment (Ref. 2-22). Furthermore, chemical reaction time scales generally are short compared with fluid motions in these combustion modes, so the many elementary reaction steps and intermediate chemical species can be neglected in this first approximation.

The concentrations of reactants and products in Equation (2-85) are usually defined by

$$-\frac{1}{2}\frac{dc_{h_2}}{dt} = -\frac{dc_{o_2}}{dt} = +\frac{1}{2}\frac{dc_{h_2o}}{dt} = \dot{\omega} \quad , \qquad (2\text{-}86)$$

where rate of reaction or reaction rate, $\dot{\omega}$, is normally proportional to the concentrations of the reactants raised to small powers that are frequently (but not necessarily) integers.

The species concentrations (moles/cm^3) are related to the macroscopic species densities through the species molecular weight

$$\rho_\alpha = c_\alpha \cdot M_\alpha \qquad (2\text{-}87)$$

and to the species volume fraction by

$$Y_\alpha = \frac{c_\alpha}{\sum_\alpha^{\alpha \neq h_2ol} c_\alpha} \quad . \qquad (2\text{-}88)$$

For the hydrogen-oxygen reaction of Equation (2-85), the reaction rate can be written

$$\dot{\omega} = k(T)\, c_{h_2}^p c_{o_2}^q \quad , \qquad (2\text{-}89)$$

where the rate constant, *k*, varies with temperature but is independent of concentration. The exponents of the concentrations in Equation (2-89) are known as the order of the reaction with respect to each reactant, where the sum of the exponents of the rate equation, i.e., order = p + q, is called the order of the reaction. There are important points concerning the order of the reaction: (1) it may be a fraction, (2) it is not necessarily related to the stoichiometric coefficients in the balance equation, and (3) it must be determined experimentally.

For this version of GASFLOW-MPI, we assume a reaction order 2, i.e., $p = 1$ and $q = 1$, which leads to the equation

$$-\frac{1}{2}\frac{dc_{h_2}}{dt} = -\frac{dc_{o_2}}{dt} = k(T)\, c_{h_2} c_{o_2} \quad . \qquad (2\text{-}90)$$

Modeling the rate constant is usually accomplished by implementing a modified Arrhenius law in the form

$$k(T)=C_f \cdot T^n \cdot e^{-\frac{E_a}{RT}} \quad , \tag{2-91}$$

where C_f is the frequency factor (we use C_f = 5x10^{12} (cm^3/mole-s-K) in this model), n is the pre-exponential temperature exponent (we use n = 0 for this model), R is the universal gas constant, and E_a is the activation energy (we use 7.8x10^{11} ergs/mole for this model) of Equation (2-85).

Using the method of partial fractions, Equation (2-90) may be analytically integrated to give the solution

$$\frac{1}{c_{h_2}(0)-2c_{o_2}(0)}\ln\left(\frac{c_{o_2}(0)}{c_{h_2}(0)}\frac{c_{h_2}(t)}{c_{o_2}(t)}\right)=k(T)\cdot t \quad . \tag{2-92}$$

Another approach is to write two individual equations for the finite-rate chemical kinetics, first for the hydrogen concentration

$$\frac{dc_{h_2}}{dt}=-2k(T)c_{h_2}c_{o_2} \tag{2-93}$$

and then the oxygen concentration

$$\frac{dc_{o_2}}{dt}=-k(T)c_{h_2}c_{o_2} \quad . \tag{2-94}$$

The chemical energy of combustion is computed as a source for the energy transport equation [Equation (2-12)] by

$$\int_V S_{I,combustion}\,dV = V\cdot C_c \cdot \dot{\omega} \quad , \tag{2-95}$$

where C_c = 4.778 x 10^{12} ergs/mole.

In practice, when solving the finite-rate chemical equations [Equations (2-93) and (2-94)] by this later method, we integrate the fuel [Equation (2-93)] when the fuel-oxidizer mixture is fuel lean and the oxidizer [Equation (2-94)] when the fuel-oxidizer mixture is fuel rich. From the chemical balance Equation (2-85), all components of the combustion process are determined.

We have compared the results of this model with the hydrogen combustion experimental data for the one-fourth-scale test facility (Ref. 2-23 and Ref. 2-24), the HDR E12 series and the Batelle Model Containment (BMC) HX series (Ref. 2-25), and an oil pool combustion test in the HDR containment building (Ref. 2-26). We have found good agreement for the general circulation patterns in complex geometries, concentrations of combustion products, and temperature distributions throughout the containment buildings. For the oil pool fires in the HDR (Ref. 2-26), the combustion model was modified to reflect hydrocarbon fuels.

2.8.2 Combustion Models Based on Reaction Progress Variable

(Warning: These combustion models are currently experimental, and should be used with caution. More details of the models will be given in the next release.)

The turbulent integral time scale, τ_t , and integral length scale , l_t , associated with the large eddies are defined as

$$\tau_t \equiv \frac{l_t}{u_t^{'}}, \tag{2-96}$$

and

$$l_t \equiv C_D \frac{\left(u_t^{'}\right)^3}{\varepsilon}, \tag{2-97}$$

where $u_t^{'} = \sqrt{\frac{2}{3}\kappa}$ is the R.M.S. turbulent velocity, ε is the turbulent dissipation rate, and $C_D = 0.37$ is the turbulent length scale constant.

The chemical time scale is defined as

$$\tau_c \equiv \frac{\alpha}{S_L^2}, \tag{2-98}$$

where α is the thermal diffusivity, and S_L is the laminar flame speed. The flame thickness, l_F, is therefore defined as

$$l_F \equiv \frac{\alpha}{S_L}. \tag{2-99}$$

Damkoehler number, *Da*, is defined as the ratio of the turbulent integral time scale to the chemical time scale

$$Da = \frac{\tau_t}{\tau_c}. \tag{2-100}$$

The progress of reaction from unburnt to burnt is represented by the scalar variable $\xi(\mathbf{x},t)$. The combustion progress variable is usually written

$$\xi(\mathbf{x},t) = \frac{Y_{H_2,initial}(\mathbf{x},t) - Y_{H_2}(\mathbf{x},t)}{Y_{H_2,initial}(\mathbf{x},t) - Y_{H_2,final}(\mathbf{x},t)}, \tag{2-101}$$

with the progress variable being either 1 (in the burnt region) or 0 (in the unburnt region). *Y* represents the mass fraction of the species.

A general "combustion progress variable" transport equation can be written as

$$\frac{\partial}{\partial t}(\rho\xi)+\nabla\cdot(\rho\xi\mathbf{u})=\nabla\cdot\left[\left(\rho\nu+\frac{\mu_t}{Sc_t}\right)\nabla\xi\right]+\rho S_\xi \quad . \tag{2-102}$$

The key to this modeling approach is the source term, ρS_ξ. Below we will outline the models for the source terms in GASFLOW-MPI.

2.8.2.1 Arrhenius Rate Model

This approach directly neglects the effect of turbulence, although indirectly turbulence is accounted for through turbulent diffusion in Equation (2-96), and assumes that chemistry plays the most important role in the combustion process. The reaction rate is given as

$$\rho S_\xi = C_f\rho(1-\xi)\exp\left(-\frac{E_a}{RT}\right) . \tag{2-103}$$

where C_f is the frequency factor (we use C_f = 5x10^{12} (cm^3/mole-s-K) in this model), and E_a is the activation energy (we use 7.8x10^{11} ergs/mole for this model).

2.8.2.2 Eddy-Break-Up Model

This model is based on phenomenological analysis of turbulent combustion for high turbulent Reynolds number (Re_t>>1) and high Damkoehler number (Da>>1). The chemical kinetic rates are neglected and the mean reaction rate is mainly controlled by turbulent mixing time, τ_t. The source term is given as

$$\rho S_\xi = -C_{EBU}\rho\frac{\varepsilon}{k}\xi(1-\xi), \tag{2-104}$$

where C_{EBU} is a model constant of the order of unity.

2.8.2.3 Eddy Dissipation Model

This model is based on the assumption that combustion occurs at small scales, where mixing occurs on a molecular level and the rate is assumed to be proportional to the inverse of the turbulent time scale. It was developed from the original eddy break-up model, the most significant difference being that the EDM model accounts for the fact that the reaction rate cannot occur unless both fuel and oxidizer mix on a molecular scale at a sufficient temperature. This is accomplished by relating the reaction rate to the limiting species. The model is formulated as follows:

$$\rho S_\xi = B_1\rho\frac{\varepsilon}{k}\min\left(Y_{H_2},\frac{Y_{O_2}}{\phi},B_2\frac{Y_{H_2O}}{1+\phi}\right), \tag{2-105}$$

where B_1 and B_2 are model constants, and ϕ is the stoichiometric oxygen to hydrogen mass ratio. When these models are used in CFD calculations, it turns out that the B_1 or B_2 need to be "tuned" within a wide range in order to obtain reasonable results for a particular problem. In GASFLOW-MPI, the default values are $B_1 = 4$ and $B_2 = 0.5$.

2.8.2.4 Models Based on Progress Variable Gradient

The source term of the mean reaction progress can be modeled as

$$\rho S_{\xi} = \rho_u S_T \left|\nabla \xi\right|, \tag{2-106}$$

where ρ_u is the density of unburnt mixture, and S_T is the turbulent flame speed. The key to this modeling approach is to find correlations for S_T. GASFLOW-MPI currently provides seven correlations for turbulent flame speed, including

$$S_T = S_L, \tag{2-107}$$

$$S_T = S_L\left(1 + \frac{\sqrt{\kappa}}{S_L}\right), \tag{2-108}$$

$$S_T = S_L\left[1 + 1.25\left(\frac{u_t^{'}}{S_L}\right)^{0.7}\right], \text{ (Kawanabe correlation)} \tag{2-109}$$

$$S_T = S_L\left\{1 - 0.195\frac{l_t}{l_F} + \left[\left(0.195\frac{l_t}{l_F}\right)^2 + 0.78\frac{u_t^{'}}{S_L}\frac{l_t}{l_F}\right]^{0.5}\right\}, \text{ (Peters correlation)} \tag{2-110}$$

$$S_T = 0.52\left(u_t^{'}\right)Da^{0.25}, \text{ (Zimont correlation)} \tag{2-111}$$

$$S_T = \left(u_t^{'}\right)^{0.75} Da^{0.25}, \text{ (Zimont-Mesheriakov correlation)} \tag{2-112}$$

$$S_T = S_L + u_t^{'}\left(\frac{Da^2}{1 + Da^2}\right)^{-0.25}. \text{ (Schmid correlation)} \tag{2-113}$$

2.8.3 Two-Step Chemical Reaction Model

(Warning: The two-step model in GASFLOW-MPI is under development as time permits.)

Another option for describing the chemical process is to use a two-step model, where the chemical reaction is divided into two parts, an induction phase and an energy release phase.

2.8.3.1 Induction Parameter Model

The induction parameter model provides a simplified approach for the inclusion of chemistry into reacting flow calculations. The primary benefit is the reduction in computation time compared with using detailed chemical mechanisms.

The chemical reaction consists of two phases, an induction phase and an energy release phase. In the first phase, intermediate species build up the radical pool, the radicals that are necessary for chain branching of the reaction. Minimal energy is released. This phase is modeled by an induction time. During the second phase, the radicals recombine, which leads to the release of the main energy of the reaction. This time is called the energy release time.

To formulate an induction parameter model, it is first necessary to determine the characteristic times of the two phases: induction and energy release times. This has been done for a wide range of initial temperatures, pressures, and gas compositions. These parameters were determined from calculations using an integration package for chemical reaction rate equations and an appropriate chemical reaction scheme.

Here, a hydrogen-air system is considered. A detailed chemical mechanism with 48 reactions and 8 reactive species, developed at the Naval Research Laboratory, was used (Ref. 2-27 and Ref. 2-28). With the GASFLOW code using a solver for stiff differential equation systems, this detailed mechanism was solved for one cell over a wide range of initial conditions. The characteristic times were determined from the temperature/time history. As a criterion for completion of the first phase of the reaction and thus the induction time, a temperature rise of 2% was chosen. At 95% of the maximum temperature, the energy release phase was considered to be finished.

Induction and energy release times were then stored in the form of a table. The initial temperature had a range from 800 to 2500 K, the pressure ranged from 0.1 to 0.6 MPa, and the hydrogen concentration ranged from 5 to 30 vol%. An interpolation routine in the code is used to retrieve the needed values from this table.

For the first phase of the reaction, a nondimensional parameter is advanced through time and the computational mesh. The equation for this induction parameter can be represented in the following form:

$$\frac{dP}{dt} = -\mathbf{u}\nabla P + \nabla(D\nabla P) + \frac{1}{\tau_{ind}} \quad . \tag{2-114}$$

The first term on the right-hand side is the convective term, the second one describes the diffusion of the parameter, and the third one is the source term.

The term *P* stands for the induction parameter, **u** for the velocity vector with which the parameter can be advected through the domain, *D* is the diffusion coefficient, and t_{ind} is the temperature-, pressure-, and composition-dependent induction time. As a diffusion coefficient, the coefficient of a typical radical, OH, was used.

The initial value of *P* is zero. For *P* smaller than 1, no energy is released and the fluid composition remains the same. When *P* reaches one, the induction time is elapsed and the energy release phase begins. We decided to treat this phase as if the energy were released stepwise linearly, depending on the energy release time and the available fuel.

Induction and energy release times are updated for each cell during every time step to reflect the actual condition the fluid is in.

2.8.3.2 Coupling of Induction Parameter Model with κ-ε Model

For calculating turbulent flames, the induction parameter model is coupled with the κ-ε model. This approach is based on the eddy-dissipation concept by Magnussen and Hjertager (Ref. 2-29) with the ignition/extinction modification introduced by Hjertager (Ref. 2-30).

Experiments have shown that the rate of combustion in flames is mainly dependent on hydrodynamic parameters. The combustion rate is limited by the rate of molecular mixing between the reactants. This mixing is linked to the rate at which turbulent eddies are dissipated. It is therefore assumed that the combustion rate is proportional to the rate of dissipation of kinetic energy of turbulence.

At first, a local turbulent Reynolds number is computed from turbulent and molecular viscosities:

$$\mathrm{Re}_t = \frac{\nu_t}{\nu} \quad . \tag{2-115}$$

If this number is smaller than a critical Re number, the energy will be released using the induction parameter model. Otherwise, the Magnussen/Hjertager model is called. The critical value is chosen on the basis of yielding reasonable results, because the approach itself is highly empirical.

Two time scales are defined. One is the turbulent eddy mixing time scale

$$\tau_t = \frac{k}{\varepsilon} \quad , \tag{2-116}$$

where κ is the turbulent kinetic energy and ε is the dissipation rate of turbulent kinetic energy. Both κ and ε are calculated in the code in the κ-ε turbulence model.

The second one is a characteristic chemical time scale which is assumed to be the induction time τ_{ind} The values for τ_{ind} are interpolated from the table.

A Damkoehler number is defined as the ratio of the chemical time to the turbulent time:

$$Da = \frac{\tau_{ind}}{\tau_t} \quad . \tag{2-117}$$

If this Damkoehler number is greater than the critical value D_{ie} , the dissipation time of the turbulent eddies is too short in comparison to the induction time, and the gas will not burn. If the number is smaller than the critical value, the combustion rate is calculated as follows:

$$\dot{\omega} = -\frac{A}{\tau_t} \rho m_{\lim} \quad , \tag{2-118}$$

where m_{lim} is the smallest of the three mass fractions, namely fuel, oxygen, or mass fraction of fuel already burnt. The constants *A* and D_{ie} are given the values 16 and 1000, respectively.

2.9 Recombiner Models

We have examined data for both the Siemens (Ref. 2-31) and NIS (Ref. 2-32) recombiner box designs. In addition, we have reviewed the analysis and modeling development of Fischer (Ref. 2-33). For the GASFLOW-MPI models of these two systems, which are shown schematically in Figure 2-1, we formulate a geometry that specifies a chimney formed by the vertical walls, and within this enclosure that is open at both the top and bottom, there is a defined reaction zone or volume. The task is to derive a model for the rate of hydrogen recombination with available oxygen in this reaction zone, given the gaseous conditions entering the recombiner at, e.g., location **IN** (shown in Figure 2-1 at the lower entrance). The model should be developed such that the projected flow area into the box can be scaled to accommodate nearly any recombiner size. We assume that the structure or the porosity for fluid flow of the recombiner members (plates or other configurations) remains uniform of any scale recombiner box. These recombiner structures also provide mass that gives a certain thermal inertia to the system.

For the Siemens recombiner box, there is considerable performance data available in Ref. 2-31, which we have condensed and presented in Figure 2-1. To understand this figure, one first computes the hydrogen volume percentage at the recombiner inlet (location **IN** in Figure 2-1). Then one reads up to the recombination rate curve and then horizontally to the left vertical axis to determine the hydrogen recombination rate in g/m^2-s (rate of hydrogen consumed per unit area inflow to the recombiner box). The recombiner efficiency for these conditions, based upon inlet and outlet hydrogen volume fractions, $Y_{h_2,\mathbf{IN}}$ and $Y_{h_2,\mathbf{OUT}}$, respectively, is determined by reading horizontally from the intersection with the hydrogen recombination rate curve to the right to intersect the efficiency curve, where the efficiency is defined as

$$\eta = 1 - Y_{h_2,\mathbf{OUT}} / Y_{h_2,\mathbf{IN}} \quad . \tag{2-119}$$

To determine the efficiency percentage as defined by Equation (2-119), one reads vertically to the top horizontal axis.

Fischer (Ref. 2-33) carefully examined the experimental data that was obtained from the Battelle Model Containment for the MC recombiner test series (Ref. 2-32), which focused on the NIS granulated recombiner box design. He found that a correlation showing the flow rate through the recombiner could be established that was dependent on the hydrogen volume fraction at the recombiner inlet and on a time constant which represented the thermal inertia of the device. In addition, Fischer found that, for steady-state operation, the volumetric flow rate was only a function of the hydrogen concentration at the recombiner inlet. Fischer also determined from the data that the efficiency for the NIS recombiner was nearly constant at 84.6% over the operating range of interest. We present Fischer's findings in Figure 2-2.

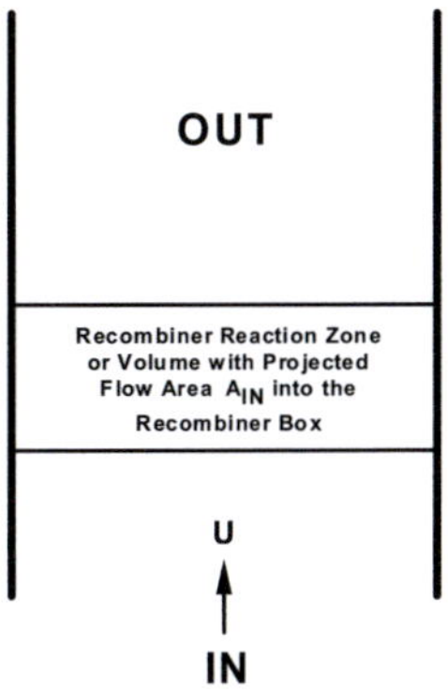

Figure 2-1 Schematic diagram for the recombiner box model

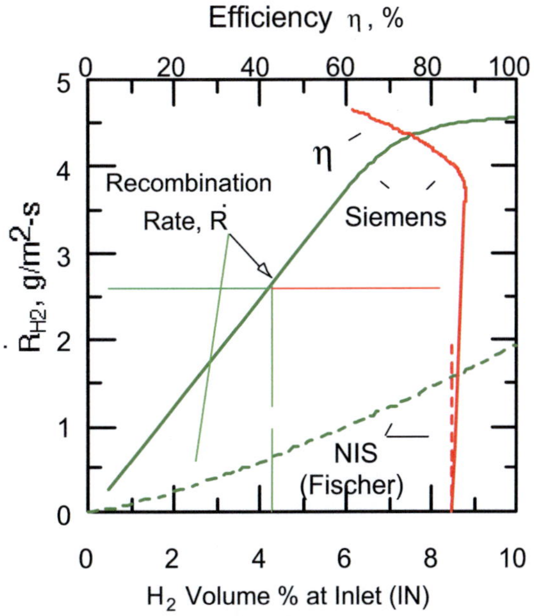

Figure 2-2 Siemens (Ref. 2-31) and NIS (Refs. 2-32 and 2-33) recombiner

In order to develop the basis for some of the GASFLOW-MPI recombiner models, we summarize Fischer's model here:

(1) The steady-state volumetric flow through the NIS recombiner, Q_0, is given by

$$Q_0 = a\left(Y_{h_2,\mathbf{IN}}\right)^b \quad , \tag{2-120}$$

where $Y_{h_2,\mathbf{IN}}$ = hydrogen volume fraction at the box inlet (**IN** in Figure 2-1)

and the experimentally determined constants are given by

a = 0.67e+06 (cm^3/s)

b = 0.307 .

(2) The time-dependent behavior of the volumetric flow *Q(t)* is described by the differential equation

$$\frac{dQ}{dt} = \frac{1}{\tau}(Q_0 - Q) \quad , \tag{2-121}$$

where τ is a relaxation time constant determined by experiment to be roughly 1800 s.

(3) By carefully examining the time-averaged data at the recombiner inlet and outlet, Fischer found that the hydrogen recombination through the flow passages was incomplete. He determined an efficiency factor based on Equation (2-119) to be η = 0.846.

2.9.1 NIS Recombiner Model

The Battelle-Frankfurt NIS tests (Ref. 2-32) and Fischer's analysis (Ref. 2-33) were conducted on a 1 m^2 cross-section inflow recombiner box of high-heat-capacity granulate design. We wish to generalize this model to be able to scale to different inflow areas but of the same or very similar internal design. In addition, we wish to derive the GASFLOW-MPI model in terms of inflow velocity, **U** in Figure 2-1, to conform with the GASFLOW-MPI numerical algorithm. Following Fischer's model, the GASFLOW-MPI model for NIS recombiner boxes proceeds as follows:

(1) The reference velocity in cm/s from Equation (2-91) is given by

$$U_0(t) = 67 \cdot \left[Y_{h_2,\mathrm{IN}}(t) \right]^{0.307} \quad . \tag{2-122}$$

Note that, in the general case, the reference inlet condition (Fischer's "steady-state value") becomes a function of time because containment conditions are nearly always changing.

(2) The time-dependent velocity, *U(t)*, inflowing into the recombiner box is

$$\frac{dU(t)}{dt} = \frac{1}{\tau}\left[U_0(t) - U(t)\right] = \frac{1}{\tau}\left\{0.67\left[Y_{h_2,\mathrm{IN}}(t)\right]^{0.307} - U(t)\right\} \quad . \tag{2-123}$$

(3) We define a recombination reaction rate as

$$\dot{R}_{h_2} = -\frac{\delta m_{h_2}}{\delta t} = -\left(\dot{m}_{h_2,out} - \dot{m}_{h_2,in}\right) = \dot{m}_{h_2,in} - \dot{m}_{h_2,out} \quad , \tag{2-124}$$

which we must relate to the efficiency based upon hydrogen volume fractions shown in Equation (2-117). With a fair amount of algebra, Equation (2-124) can be reformulated to

$$\dot{R}_{h_2} = \eta \cdot \dot{m}_{h_2,in} \left[\frac{1}{1 - \frac{1}{2}(1-\eta) \cdot Y_{h_2,in}} \right] \quad , \tag{2-125}$$

which reflects the fact that the number of moles flowing through the recombiner changes during the recombination process. For the conditions we are interested in, see Figure 2-2, the term in brackets is very close to unity, so we can simply write the hydrogen recombination rate as

$$\dot{R}_{h_2} = \eta \cdot \dot{m}_{h_2,in} \quad . \tag{2-126}$$

We now relate the hydrogen recombination rate from Equation (2-126) to the reaction rate from Equation (2-86) by

$$\dot{R}_{h_2} = 2 \cdot \dot{\omega} \cdot V \cdot M_{h_2} \quad , \tag{2-127}$$

where $\dot{\omega}$ is the rate of hydrogen recombination in moles-h2/cm^3-s . Knowing the efficiency is 0.846, we can compute the rate of recombination, $\dot{\omega}$, as

$$\dot{\omega} = \frac{1}{2} \frac{0.846 \cdot U(t) \cdot A_{\mathrm{IN}} \cdot c_{h_2,\mathrm{IN}}(t)}{V} \quad . \tag{2-128}$$

(4) We use this recombination rate to solve the chemical kinetics, Equation (2-86), for the consumption of hydrogen and oxygen and the production of water vapor, as well as the combustion energy source term

$$\int_V S_{I,recombination} dV = V \cdot C_c \cdot \dot{\omega} \tag{2-129}$$

released in the recombiner reaction zone. Since Equation (2-128) is computed to be a constant based on conditions and properties at the recombiner inlet, careful evaluation of the chemical kinetics, Equation (2-86), must follow to ensure non-negative species concentrations.

2.9.2 Siemens Recombiner Model

We have developed a recombiner model for the Siemens recombiner box in a similar fashion. The GASFLOW-MPI model for the Siemens recombiner is as follows:

(1) We first calculate the time-dependent hydrogen and oxygen volume fractions, $Y_{h_2,IN}(t)$ and $Y_{o_2,IN}(t)$, respectively, at the recombiner inlet (location **IN** in Figure 2-1).

(2) A check of the species composition is made from

$$Y_{IN}(t) = \min\left[Y_{h_2,IN}(t), 2\frac{Y_{o_2,IN}(t)}{\eta(t)} \right] \quad . \tag{2-130}$$

(3) Using the volume fraction as a volume percentage, one then determines both the recombination rate, $\dot{R}(t)$, and recombination efficiency, $\eta(t)$, from Figure 3-1.

(4) A reference velocity is computed as follows:

$$U_0(t)=\frac{\dot{R}(t)}{A_{\mathbf{IN}}\cdot\eta(t)\cdot\rho_{h_2,\mathbf{IN}}(t)} \quad , \tag{2-131}$$

where $A_{\mathbf{IN}}$ is the inflow area into the recombiner box in m^2.

(5) The time-dependent velocity, *U(t)*, inflowing into the recombiner box is

$$\frac{dU(t)}{dt}=\frac{1}{\tau}\left[U_0(t)-U(t)\right]=\frac{1}{\tau}\left\{\frac{\dot{R}(t)}{A_{\mathbf{IN}}\cdot\eta(t)\cdot\rho_{h_2,\mathbf{IN}}(t)}-U(t)\right\} \quad , \tag{2-132}$$

where τ for this model is 10 s.

(6) The chemical kinetics with change of species concentrations, Equation (2-86), is computed from

$$\dot{\omega}=\frac{1}{2}\frac{\eta\cdot U(t)\cdot A_{\mathbf{IN}}\cdot c_{h_2,\mathbf{IN}}(t)}{V} \quad , \tag{2-133}$$

and the energy release, Equation (2-126), is updated in the recombiner reaction zone.

Again, care must be taken to insure positive species concentrations with this model because the chemical kinetics in the recombiner reaction zone is dependent on the inflow conditions and properties of the recombiner box.

2.9.3 Siemens Correlation for Siemens Type FR-90/1 Recombiner

The Siemens recombiner correlation (Ref. 2-34) can be used for a Siemens recombiner of the type FR-90/1. This relationship between the species volume fractions, Y_{h_2} and Y_{o_2}, and the pressure *p* in bars is given by

$$\int_V S_{p,h_2,Siemens\_correlation}\,dV=-\begin{Bmatrix} 0;\begin{Bmatrix} Y_{h_2}<0.02 \\ or \\ Y_{o_2}<0.0005 \end{Bmatrix};otherwise \\ \min\left(Y_{h_2},2\cdot Y_{o_2},0.08\right)\cdot\left(k_1\cdot p+k_2\right) \end{Bmatrix} \quad , \tag{2-134}$$

where the empirically determined constants, k_1 and k_2, are given in Tabele 2-5.

Table 2-5 Siemens recombiner constants of FR-90/1

Type	k_1 (g/s-bar)	k_2 (g/s)
FR-90/1-320	0.010	0.012
FR-90/1-960	0.031	0.037
FR-90/1-1500	0.137	0.167

In order to utilize this correlation in the GASFLOW-MPI methodology, we recast the correlation into

$$\dot{\omega}=\frac{1}{2}\frac{\left\{\begin{array}{l} 0;\left\{\begin{array}{l} Y_{h_2}<0.02 \\ or \\ Y_{o_2}<0.0005 \end{array}\right\};otherwise \\ \min\left(Y_{h_2},2\cdot Y_{o_2},0.08\right)\cdot\left(k_1\cdot p+k_2\right) \end{array}\right\}}{M_{h_2}\cdot V} \quad . \tag{2-135}$$

The chemical kinetics can be evaluated when Equations (2-86) and (2-135) are integrated. Computing the energy from the recombination process with Equation (2-126) completes this model.

2.9.4 GRS Recombiner Correlation

The GRS recombiner correlation model (Ref. 2-35) can be used for a single recombiner plate or foil. This correlation is of an Arrhenius type,

$$\int_V S_{\rho,h_2,GRS\_correlation} dV=-Area\cdot k_1\cdot\left(c_{h_2}\right)^{1.05}\cdot e^{-\Delta E(T)/(\Re\cdot T)} \quad , \tag{2-136}$$

where *Area* is the recombiner surface area, k_1 is a constant equaling 5.0x10^2, c_{h_2} is the hydrogen concentration, $\Re$ is the universal gas constant, *T* is the temperature, and $\Delta E(T)$ is an activation function approximated by

$$\Delta E(T)=\left\{\begin{array}{l} 20.271+\left[2.1771\cdot 10^{-2}\left(\begin{array}{l} -1.5078\cdot 10^{-5}- \\ 9.4707\cdot 10^{-10}\cdot T \end{array}\right)\cdot T\right]\cdot T;T<506.9K \\ 1.8322+5.026\cdot 10^{-2}\cdot T;T\geq 506.9K \end{array}\right. \quad . \tag{2-137}$$

To cast this relationship into the GASFLOW-MPI methodology, we rearrange this expression to

$$\dot{\omega}=\frac{1}{2}\frac{Area\cdot k_1\cdot\left(c_{h_2}\right)^{1.05}\cdot e^{-\Delta E(T)/(\Re\cdot T)}}{M_{h_2}\cdot V} \quad . \tag{2-138}$$

The chemical kinetics can be evaluated when Equations (2-86) and (2-138) are integrated. Computing the energy from the recombination process with Equation (2-126) completes this model.

2.10 Ignition Model

The ignitor model is simple, but effective. For a user-specified location, the gaseous composition is first checked to see if a combustible mixture of hydrogen and oxygen exists. We essentially implement the lean combustion limit from the Shapiro Diagram to evaluate the threshold for which a mixture of hydrogen, oxygen, and steam is flammable. The idea is that the hydrogen volume percentage must exceed 4% for steam volume percentages up to 30% with increasing hydrogen volume percentages to 12% for steam volume percentages to 65%. Above 65% steam volume percent, the mixture is considered to be steam inerted. This criteria can be written as

$$\begin{cases} h_2o\left\langle Volume\%_{h_2o}\right\rangle < 65 \\ h_2\left\langle Volume\%_{h_2o}\right\rangle \geq 4+\min\left[0,\frac{8}{35}\left(Volume\%_{h_2o}-30\right)\right] \end{cases} . \tag{2-139}$$

If this condition is met, the temperature in the reaction rate constant of Equation (2-90) is set equal to 2000 K,

$$k\left(2000\right)=C_f\cdot e^{-\frac{E}{\Re\cdot 2000}} , \tag{2-140}$$

and the chemical kinetics equations are solved with this driving function.

We are able to model both a "glow plug" and a "spark" type of ignitor. With a glow-plug-type ignitor, the reaction rate constant , Equation (2-137), is continuously active for the specified time interval, whereas for the spark type ignitor, the reaction rate constant is only active at a specified spark interval (sparking frequency) and specified sparking duration.

2.11 Lagrangian Discrete Particle Model

The GASFLOW-MPI aerosol model comprises a Lagrangian discrete particle transport model, a stochastic turbulent particle diffusion model, a particle deposition model, a particle entrainment model, and a particle cloud model. These models incorporate the physics of particle behavior to model discrete particle phenomena and allow the code user to track the transport, and deposition and entrainment of discrete particles, as well as, clouds of particles in nuclear systems.

2.11.1 Particle Transport

The inertial force, $\mathbf{F}_I$, of a particle is equal to the sum of the external aerodynamic forces acting on the particle. The external forces modeled are aerodynamic drag , $\mathbf{F}_d$, gravitational, $\mathbf{F}_{gr}$, and centrifugal, $\mathbf{F}_C$. The force balance equation for each particle is

$$\mathbf{F}_I = \mathbf{F}_d + \mathbf{F}_{gr} + \mathbf{F}_c = \frac{\pi}{6} d_p^3 \rho_p \frac{d\mathbf{U}_p}{dt}, \tag{2-141}$$

where d_p is particle diameter, r_p is particle density, $\mathbf{U}_p$ is particle velocity, and t is time.

Convection is the primary means of particle transport. The particle velocity is nearly that of the conveying fluid, but this small difference in velocity gives rise to the primary aerodynamic force, the drag force, acting on particles in motion,

$$\mathbf{F}_d = C_D \frac{\pi}{8} d_p^2 \rho_g \left(\mathbf{U}_g - \mathbf{U}_p\right)^2, \tag{2-142}$$

where C_D is the particle drag coefficient, ρ_g is gas density, and $\mathbf{U}_g$ is the gas velocity at the particle location. The fluid drag force is computed from Newton's resistance law to determine the form drag for $\mathrm{Re}_p > 906.291$. In this case, the drag coefficient is $C_D = 0.44$. The influence of the Reynolds number on the drag coefficient of spherical particles over a wide range is represented as

$$C_d = \begin{cases} \dfrac{24}{\mathrm{Re}_p}\left(1 + \dfrac{\mathrm{Re}_p^{\ 2/3}}{6}\right); & \mathrm{Re}_p < 906.291 \\ 0.44 & ;\ otherwise \end{cases}, \tag{2-143}$$

(see Ref. 2-37). This formulation is used in the code.

The force from gravity is

$$\mathbf{F}_{gr} = \frac{\pi}{6} d_p^3 \mathbf{g}(\rho_p - \rho_g), \tag{2-144}$$

where $\mathbf{g}$ is gravitational acceleration. When a particle is released in air it quickly reaches a terminal velocity, a condition in which the drag force of the air will be exactly equal and opposite to the force of gravity.

A particle is said to have curvilinear motion when it follows a curved path rather than a straight-line or oscillatory motion. A particle traveling along a curved streamline in flow around an obstacle will have curvilinear motion and will experience a centrifugal force, which can be expressed as

$$\mathbf{F}_c = \frac{\pi}{6} d_p^3 \rho_p \frac{\mathbf{U}_t^2}{R}, \tag{2-145}$$

where $\mathbf{U}_t$ is the tangential velocity at radius of curvature R. Curvilinear motion is characterized by the dimensionless Stokes number (S_{tk}), which is the ratio of the stopping distance (a measure of the inertial range of a particle) to a characteristic dimension of the obstacle. For geometrically similar particle motion to occur around differently sized obstacles, the flow Reynolds numbers must be equal and the Stokes numbers must be equal.

2.11.2 Particle Turbulent Diffusion

The fluid velocity at the location of a discrete particle is the sum of the mean velocity components and the turbulent velocity components. The turbulent velocity is determined by a stochastic approach suggested by Hotchkiss and Hirt (Ref. 2-38). The idea is to consider the particle as a point source that diffuses for a time Δt. The probability of where the particle is likely to move along a coordinate axis x can be written in the form of a Gaussian probability distribution function

$$n(x,t) = \frac{1}{\sqrt{2\pi\sigma}} e^{\frac{-x^2}{2\sigma^2}} \quad (2\text{-}146)$$

where $2\sigma^2$ is the dispersion of the probability distribution, the standard deviation $\sigma = \sqrt{2D_p \Delta t}$, and D_p in the standard deviation equation, is the turbulent diffusion coefficient. A random number generator selects the actual location used within the possible distribution. The turbulent velocity component is determined from the randomly chosen location. This process is repeated for each of the three coordinate directions.

2.11.3 Particle Deposition

Particle adhesion is poorly understood and its description is partly qualitative. Hinds states that because it is such a complicated phenomenon there is no complete theory that accounts for all the factors that influence adhesion (Ref. 2-39). The main adhesive forces are the van der Waals force, electrostatic forces, and the surface tension of absorbed liquid films. These forces are affected by the following: the material, shape, and size of the particle; the material, roughness, and contamination of the surface; the relative humidity and temperature of the ambient gas; and the duration of particle-surface contact and initial contact velocity.

The theoretical analyses of adhesion forces presented by Hinds (Ref. 2-39) and Dahneke (Ref. 2-40 and Ref. 2-41) indicate that the most important adhesion forces are the London-van der Waals forces, the long-range attractive forces that exist between molecules. Hinds explains that these forces arise because the random movement of electrons in any material creates momentary areas of charge concentration called dipoles. At any instant these dipoles induce complementary dipoles in neighboring material, which gives rise to attractive forces. These forces decrease rapidly with separation distance between surfaces; consequently, their influence extends only several molecular diameters away from a surface.

Small particles do not always adhere to a surface after impingement. At "low" velocities a particle may lose all of its kinetic energy on impact by deforming itself and the surface. In this velocity range, the greater the deformation the greater the adhesive force. At "high" velocities, part of the particle kinetic energy is dissipated in the deformation process and part is converted elastically to kinetic energy of rebound. At some threshold velocity the rebound energy will exceed the adhesion energy and the particle will bounce off the surface. This can occur for particle sizes that would adhere tightly in a static situation. The harder the particle and surface, the larger the particle, and the greater the velocity, the more likely bounce is to occur.

To model particle deposition in GASFLOW-MPI, the theory of rebound developed by Dahneke (Ref. 2-40 and Ref. 2-41) is used. Dahneke's rebound theory includes an energy balance analysis of the impinging and rebound particle. Consider a particle moving towards a surface, still beyond influence of the surface, with incident normal velocity $\mathbf{U}_i$. The particle kinetic energy due to $\mathbf{U}_i$ is KE_i. (Kinetic energy due to motion parallel to the surface is assumed conserved throughout the collision, and kinetic energy due to spinning of the particle is neglected.) Upon nearing the surface the particle will fall into the particle-surface potential energy well of depth E. In general, the depth of the potential well may vary during the collision. Possible causes could be contact charging or particle and surface deformation. Thus two potential well depths are defined: E_i is the depth seen by the incident particle, and E_r is the depth seen by the reflected particle. To obtain the kinetic energy of the particle at the instant of rebound, the concept of the coefficient of restitution, e, is used. This is defined as the ratio of normal particle velocity at the instant of rebound to normal particle velocity at the instant of contact. The kinetic energy at the instant of rebound is

$$KE_{at\ rebound} = (KE_i + E_i)e^2, \tag{2-147}$$

where the sum of KE_i and E_i is the kinetic energy of the particle at the instant of contact, and e^2 is the fraction of this energy recovered by the particle. If the collision is perfectly elastic, $e^2 = 1$.

Upon rebound, the particle must exchange kinetic energy for potential energy as it climbs out of the potential energy well. The final kinetic energy of the reflected particle, beyond influence of the surface, is

$$KE_r = (KE_i + E_i)e^2 - E_r. \tag{2-148}$$

In terms of particle velocity, $\mathbf{U}$, and mass, m, Equation (2-148) can be equivalently written as

$$\frac{\mathbf{U}_r}{\mathbf{U}_i} = \left[e^2 + \frac{2\left(E_i e^2 - E_r\right)}{m\mathbf{U}_i^2} \right]^{\frac{1}{2}}, \tag{2-149}$$

where U_r is the rebound velocity. Dahneke (Ref. 2-40) noted that if one could approximate $E = E_r = E_i$, then Equation (2-149) would simplify to

$$\frac{\mathbf{U}_r}{\mathbf{U}_i} = \left[e^2 + \frac{2E\left(e^2 - 1\right)}{m\mathbf{U}_i^2} \right]^{\frac{1}{2}}. \tag{2-150}$$

When there is no rebound, the particle must fail to climb out of the potential energy well, and thus KE_r is zero, and

$$\mathbf{U}_i^* = \left[\frac{2E}{m}\left(\frac{1-e^2}{e^2} \right) \right]^{\frac{1}{2}}. \tag{2-151}$$

This equation determines the critical rebound speed, $\mathbf{U}_i^*$; it describes the limiting case between particle capture and rebound.

The depth of the potential energy well for a sphere of diameter *d* and a flat surface adhering to each other is given by Dahneke (Ref. 2-40), quoting as references Bradley (Ref. 2-42) and Hamaker (Ref. 2-43), as

$$E = \frac{Ad}{12z_0}, \tag{2-152}$$

where *A* is the Hamaker constant, z_0 is the equilibrium separation of the sphere and surface, and *d* is the particle diameter.

The attractive force between the sphere and a surface is

$$\mathbf{F} = -dE / dz = -\frac{Ad}{12z_0^2} \quad . \tag{2-153}$$

Combining Equations (2-151) and (2-152) and writing the particle mass in terms of its volume and density, ρ_p, gives

$$\mathbf{U}_i^* = \left[\frac{A\left(1-e^2\right)}{\pi z_0 \rho_p d^2 e^2}\right]^{\frac{1}{2}} \quad . \tag{2-154}$$

Dahneke (Ref. 2-41) states that the principal deficiency of the Bradley-Hamaker derivation of the adhesion energy is its failure to consider the repulsive portion of the molecular interactions. Thus, flattening of the sphere tip and deformation of the surface are not considered. Dahneke presents in the 1972 publication (Ref. 2-41) a new theory for the adhesion of particles (spheres) that includes the influence of the repulsive component of force from the elastic flattening of the spheres. The repulsive component of force is expressed in terms of the apparent penetration depth, *h*, of the impacting sphere into the substrate; the sphere size, *d*; and the sphere bulk mechanical properties, $K = (1 - N^2) / Y$, where *N* is Poisson's ratio and *Y* is Young's modulus of the particle material. The net force between two impacting spheres (or a sphere and a surface) is written as the sum of the attractive and repulsive forces:

$$\mathbf{F} = \left(\frac{Ad}{12z_0^2}\right)\left(1+\frac{2h}{z_0}\right)+\left(\frac{2\sqrt{2d}}{3K}\right)h^{\frac{3}{2}} \quad . \tag{2-155}$$

In order to derive an equation for the maximum adhesive force, $\mathbf{F}_{max}$, Dahneke assumes that some flattening will always occur when two spheres (or a sphere and a flat surface) adhere and exploits the fact that **F** must obtain a minimum at some value of penetration *h*. He states that the expression for $\mathbf{F}_{max}$ is of practical importance in calculating adhesive force, since a force greater than $\mathbf{F}_{max}$ is required to separate two spheres or a sphere from a surface by, for example, centrifuging or blowing, or by dislodgment through inertial forces when a stress in the surface material is reflected to the surface. The maximum adhesive force is

$$\mathbf{F}_a = \left(\frac{Ad}{12z_0^2}\right)\left[1+\left(\frac{A^2K^2}{108z_0^7}\right)d\right] , \tag{2-156}$$

which occurs at the penetration

$$h_f = \left(\frac{A^2K^2}{72z_0^6}\right)d \quad . \tag{2-157}$$

Note that the first term in Equation (2-156) is the adhesive force between two unflattened spheres. The second term is the increase in adhesive force due to flattening of the spheres. Thus, flattening has greater effect for larger particle sizes and softer materials.

In order to obtain an expression for the threshold bounce velocity, $\mathbf{U}_i^*$, including the effect due to flattening, it is first necessary to obtain an expression for the interaction energy. This is obtained directly from Equation (2-155) using the relationship

$$E = \int \mathbf{F} dh .$$

This yields

$$E = \left(\frac{Ad}{12z_0}\right)\left(1+\frac{h}{z_0}+\frac{h^2}{z_0^2}\right)+\left(\frac{4\sqrt{2d}}{15K}\right)h^{\frac{5}{2}} . \tag{2-158}$$

The potential energy well equivalent to the energy required to overcome the maximum adhesive force expressed in Equation (2-156) is obtained by substituting the penetration depth, h_f, expressed in Equation (2-157) into Equation (2-158). This gives

$$E = \left(\frac{Ad}{12z_0}\right)\left[1+\left(\frac{A^2K^2d}{72z_0^7}\right)+\left(\frac{A^4K^4d^2}{72^2z_0^{14}}\right)\right]+\left(\frac{4\sqrt{2d}}{15K}\right)\left(\frac{A^2K^2d}{72z_0^6}\right)^{\frac{5}{2}} \quad . \tag{2-159}$$

Combining Equations (2-151) and (2-159) gives the equation for the critical rebound velocity that determines the threshold particle bounce velocity.

The critical rebound velocity is the value of velocity for a 50% probability of bounce. The concept of the threshold velocity or critical rebound velocity $\mathbf{U}_i^*$ being defined as 50% likelihood of particle bounce is presented by Hinds (Ref. 2-39) and Paw U (Ref. 2-44). The incident velocity window outside of which the particle either adheres or bounces is somewhat arbitrary. Data presented by Paw U suggest that plus or minus 50% of $\mathbf{U}_i^*$ is a reasonable assumption for this. That is, for $\mathbf{U}_i < \mathbf{U}_i^*(0.5)$ the particle always adheres, and for $\mathbf{U}_i > \mathbf{U}_i^*(1.5)$ the particle always bounces. With these specified limits the probability of particle bounce for the velocity range $0.5\,\mathbf{U}_i^* < \mathbf{U}_i < 1.5\,\mathbf{U}_i^*$ is expressed as

$$0 \le \left[\frac{\left(-0.5\mathbf{U}_i^* + \mathbf{U}_i\right)}{\mathbf{U}_i^*}\right] \le 1 \quad . \tag{2-160}$$

A determination of whether or not specific particles will bounce or adhere can be made by generating a random number, α with a value between 0 and 1, and testing for

$$\alpha \le \left[\frac{\left(-0.5\mathbf{U}_i^* + \mathbf{U}_i\right)}{\mathbf{U}_i^*} \right] \quad . \tag{2-161}$$

When this condition is met, the particle will bounce.

Dahneke's data (Ref. 2-45 and Ref. 2-46) show the trend for the coefficient of restitution, *e*, to reach a maximum value, in this case 0.96, at the threshold bounce velocity of about 10 cm/s. Almost immediately the ratio of rebound velocity to incident velocity begins to decrease as the incident velocity increases. At an incident velocity of 100 cm/s, *e* is 0.68; at 400 cm/s, *e* is 0.15. An approximately 70% decrease in *e* results in an increase in incident velocity of 100 cm/s. The following equation models this data relatively well:

$$e = e_0 \left(0.60\right)^{\exp} \quad , \tag{2-162}$$

where

$$\exp = \frac{\mathbf{U}_i - \mathbf{U}_i^*}{10.0\mathbf{U}_i^*} \quad ,$$

and e_0 is the coefficient of restitution at the threshold bounce velocity.

Particle deposition and bounce are not understood well enough that a totally mechanistic model can be developed. The phenomena simulated in this deposition model are based on theoretical models that are necessarily restricted by a wide range of assumptions and experimental data that are limited to specific conditions. Deposition is in a real sense a stochastic process that follows the general trend of the theoretical and empirical models developed and compared with available experimental data. Because of this, it is a reasonable assumption that some small, unknown percentage of the particles that impact a surface will adhere. To account for this, currently 5% of all particles that impact a surface do adhere. The particles that adhere are randomly chosen.

2.11.4 Particle Entrainment

The mechanisms involved in the entrainment of deposited particles from a surface into the suspension fluid are not well understood. Suspension is associated with the aerodynamic detachment of particles for which the principal force holding them onto a surface is the intersurface molecular force (adhesion force).

Suspension is initiated by flows when the fluid velocity equals or exceeds the particle threshold suspension velocity value. As pointed out by Halow (Ref. 2-47), most treatments of particle suspension have provided correlation of a largely empirical nature and, while perhaps describing the available experimental data, do not have a sound enough fundamental basis to permit confident extrapolation of the data to uninvestigated systems and conditions. Most theoretical work has considered the balance of forces acting on a suspension particle, considering the fluid-particle interaction in the shear flows in the viscous sublayer, which are assumed to be steady. A force balance approach modified by experimental data is used in this model.

Criteria for determining the fluid velocity at which a particle initially at rest on a surface will become suspended can be determined from a force balance equation that includes gravity, adhesion, fluid lift, fluid drag, and friction forces. A particle adhering to a surface will be dislodged when the removal forces *equal or exceed* the force of particle adhesion. This model is based on a force balance approach plus experimental data reported by Cabrejos and Klinzing.* They present a technique for finding the minimum pickup (suspension) velocity of solid particles in horizontal pneumatic conveying. This general semi-empirical correlation is based on the Archimedes number and is valid over a range of particle sizes from 10 to 1000 mm.

When the size of the particle is smaller than the thickness of the viscous sub-layer, that is, $d_p < d_l$, the forces acting on the sphere are as shown in Figure 2-3, where a horizontal flow is assumed.

Gravitational, buoyant, and adhesive forces depend only on the physical properties of the particle and the gas density, and they are independent of the gas stream velocity. The drag and lift forces depend on the gas stream velocity, whereas the frictional force is proportional to the coefficient of sliding friction. The adhesive force is the van der Waals intersurface molecular force.

The forces acting on a small sphere, immersed in the viscous sublayer are as follows:

(1) Gravitational force: $\mathbf{F}_g = \frac{\pi}{6} d_p^3 \rho_p \mathbf{g}$,

(2) Buoyant force: $\mathbf{F}_b = \frac{\pi}{6} d_p^3 \rho_g \mathbf{g}$,

(3) Adhesive force: $\mathbf{F}_a = \left(\frac{A d_p}{12 z_0^2}\right)\left[1 + \left(\frac{A^2 K^2}{108 z_0^7}\right) d_p\right]$,

(4) Drag force: $\mathbf{F}_D = C_D \frac{\pi}{8} d_p^2 \rho_g \mathbf{U}_{gcp}^2$,

(5) Lift force: $\mathbf{F}_l = 6.46 \rho_g \sqrt{\nu} \mathbf{U}_{gcp} \sqrt{\frac{\partial \mathbf{U}_g}{\partial y} \frac{d_p^2}{4}}$, and

(6) Friction force: $\mathbf{F}_f = f_s \mathbf{F}_n = f_s \left(\mathbf{F}_g + \mathbf{F}_a - \mathbf{F}_b - \mathbf{F}_l\right)$

where $\mathbf{U}gcp$ is the gas stream velocity at the center ($y = d_p/2$) of the stationary particle, which is typically in the viscous sublayer. The shear velocity, U_t, is known, so $\mathbf{U}_{gcp}$ can be computed from

* Francisco J. Cabrejos and George E. Klinzing, "Incipient Motion of Solid Particles in Horizontal Pneumatic Conveying," Dept. of Mech. Engr., University of Pittsburg, unpublished paper (1991).

$$\mathbf{U}_{gcp} = \frac{d_p U_\tau^2}{2\nu}. \tag{2-163}$$

The adhesive and lift forces are from Dahneke (Ref. 2-41) and Saffman (Ref. 2-48), respectively.

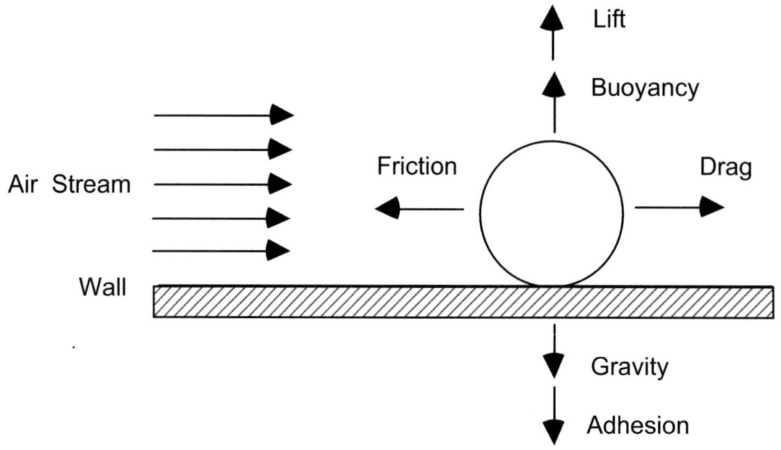

Figure 2-3 Forces acting on a single sphere at rest on a wall with a steady, fully developed turbulent flow

The small sphere will begin moving when the forces acting on the particle are zero. Applying Newton's second law, motion will take place when

$$\mathbf{F}_d = f_s\left(\mathbf{F}_g + \mathbf{F}_a - \mathbf{F}_b - \mathbf{F}_l\right). \tag{2-164}$$

Substituting the above force equations into the force balance equation gives

$$C_D \frac{\pi}{8} d_p^2 \rho_g \mathbf{U}_{gcp}^2 / f_s = \frac{\pi}{6} d_p^3 \left(\rho_p - \rho_g\right)\mathbf{g} + \left(\frac{A d_p}{12 z_0^2}\right)\left[1 + \left(\frac{A^2 K^2}{108 z_0^7}\right) d_p\right]$$

$$-6.46 \rho_g \sqrt{\nu}\,\mathbf{U}_{gcp} \sqrt{\frac{\partial \mathbf{U}_g}{\partial y} \frac{d_p^2}{4}}. \tag{2-165}$$

A Newton-Raphson solver can be used to solve this implicit equation for $\mathbf{U}_{gcp}$, which is the minimum pickup velocity for the single particle.

This model follows the derivation of the force balance model presented by Cabrejos and Klinzing.* However, they derived the equations for pipe flow. They also used Munroe's equation to model the effect that the pipe would have on large particles and used Blasius' correlation for the friction factor inside a smooth pipe for the effect the pipe would have on small particles, that is, for those totally immersed in the viscous sublayer. The equations presented here are for flow over a flat plate.

---

* Francisco J. Cabrejos and George E. Klinzing, "Incipient Motion of Solid Particles in Horizontal Pneumatic Conveying," Dept. of Mech. Engr., University of Pittsburg, unpublished paper (1991).

The shear velocity, U_t, is used to determine the velocity at the small particle center in the viscous sublayer. Cabrejos and Klinzing modified the pickup velocity predicted by these fundamental equations with experimental data. Although this data was obtained in pipe flow, we believe it can be used for a more general flow model. Comparison of minimum pickup velocities predicted by the modified Cabrejos-Klinzing model with experimental measurements of the threshold suspension velocity, for flow over thick beds of a wide range of material sizes and densities (Ref. 2-49), shows good agreement. The predicted pickup velocity is modified by an empirical equation that is a function of the Archimedes number,

$$Ar = \frac{\mathbf{g}}{\nu^2}\frac{(\rho_p - \rho_g)}{\rho_g} d_p^3 . \tag{2-166}$$

The magnitude of this correction factor ranges from about 2.5 for high-density, large particles to about 250 for low-density, small particles. Cabrejos and Klinzing show that this semi-empirical model agrees well with experimental data.

Combining the single particle model, which predicts a pickup velocity, $\mathbf{U}_{gpu0}$, with the experimental data for the minimum pickup velocity of a layer of particles gives a general correlation. This semi-empirical correlation is believed to be valid over a range of particle sizes from 10 to 1000 mm. It is expressed as

$$\mathbf{U}_{gpu} = \left(1.27 Ar^{-\frac{1}{3}} + 0.036 Ar^{\frac{1}{3}} + 0.45\right)\left(0.70 Ar^{-\frac{1}{5}} + 1.0\right)\mathbf{U}_{gpu0} . \tag{2-167}$$

The first term on the right-hand side takes into account the effects of particle interactions with the other particles, and the second term accounts for the particle shape.

2.11.5 Boundary Layer Thickness

The boundary layer thickness, d_l, is approximated by

$$\frac{\delta_l}{x} = 0.370 R_x^{-\frac{1}{5}} , \tag{2-168}$$

where $R_x = u_y x/n$, and x is the distance along the wall from the point at which the turbulent fluid initially contacts the wall. To estimate the magnitude of the boundary layer thickness, assume a 300 cfm flow through a 2 ft by 2 ft duct, which is a flow velocity of about 40 cm/s. Then, 10 m from the duct entrance (x =10 m) the boundary layer thickness, d_l, is about 30 cm. Since the inner layer is approximately 2% of d_l, the inner layer is approximately 0.6 cm or 6000 mm. For a flow velocity of only 10 cm/s and at a distance of 1.0 m from the duct entrance, the boundary layer thickness is only 5 cm, and the inner layer is 1000 mm. These examples indicate that generally the deposited particles available for resuspension will be immersed in the inner layer of the boundary layer.

2.11.6 Particle Cloud Model

The particle cloud model permits each discrete computational particle to represent a cluster of particles that can disperse as a Gaussian cloud. The density and size of the particles in the cloud of particles are the same as the computational particle with which the cloud is associated. The particle cloud density, ρ_{pc}, at a selected point that is a distance *r* from the cloud center is given by

$$\rho_{pc}(r,t) = M_{pc} f(r,\sigma), \tag{2-169}$$

where

$$f(r,\sigma) = \left(\frac{1}{(2\pi)^{\frac{3}{2}}\sigma^3}\right) e^{\frac{-r^2}{2\sigma^2}} \tag{2-170}$$

and

$$\sigma = r_{pci} + \sqrt{2D_{pc}t}\,. \tag{2-171}$$

The term M_{pc} is the mass of the particle cloud, D_{pc} is the particle diffusion coefficient of the cloud, r_{pci} is the initial radius of the particle cloud, and *t* is the elapsed cloud growth time. The radius of the cloud at any elapsed time is r_{pc} = 3 s. The cloud density at any given point is determined by the summation of all density contributions of individual particle clouds at that point.

References

Ref. 2-1. G. K. Batchelor, An Introduction to Fluid Dynamics (Cambridge University Press, Cambridge, 1967).

Ref. 2-2. P. A. Thompson, Compressible-Fluid Dynamics (McGraw-Hill Book Co., New York, 1972).

Ref. 2-3. C. W. Hirt, "Volume-Fraction Techniques: Powerful Tools for Wind Engineering," Journal of Wind Engineering and Industrial Aerodynamics, 46 & 47, 327-338 (1993).

Ref. 2-4. R. B. Bird, W. E. Stewart, and E. N. Lightfoot, Transport Phenomena,(John Wiley and Sons, New York, 1960).

Ref. 2-5. S. R. Turns, An Introduction to Combustion,(McGraw-Hill, New York, 1996).

Ref. 2-6. J. D. Ramshaw, "Self-Consistent Effective Binary Diffusion in Multicomponent Gas Mixtures," J. Non-Equilib. Thermodyn., 15, 295-300 (1990).

Ref. 2-7. JANAF Thermochemical Tables, Third Edition, National Bureau of Standards, New York (1985) .

Ref. 2-8. S. Gordon and B. J. McBride, "Computer Program for Calculation of Complex Chemical Equilibrium Compositions, Rocket Performance, Incident and Reflected Shocks, and Chapman-Jouget Detonations," NASA Special Publication SP-273 (1971).

Ref. 2-9. JANAF Thermochemical Tables, Second Edition, National Bureau of Standards, New York (1971) .

Ref. 2-10. Keenam, Keyes, Hill, and Moore, Steam Tables (John Wiley and Sons, New York, 1969).

Ref. 2-11. R. J. Kee, J. A. Miller, and T. H. Jefferson, "CHEMKIN: A General-Purpose, Problem-Independent, Transportable, Fortran Chemical Kinetics Code Package," Sandia National Laboratories Report, SAND80-8003 (1980).

Ref. 2-12. S. Mathur, P. K. Tondon, and S. C. Saxena, "Thermal Conductivity of Binary, Ternary and Quaternary Mixtures of Rare Gases," Mol. Phys., 12, 569 (1967).

Ref. 2-13. C. R. Wilke, "A Viscosity Equation for Gas Mixtures," J. Chem. Phys., 18, 517 (1950).

Ref. 2-14. C. F. Curtiss and J. O. Hirschfelder, "Transport Properties of Multicomponent Gas Mixtures," J. Chem. Phys., 17, 550 (1949).

Ref. 2-15. B. E. Launder and D. B. Spalding, "The Numerical Computation of Turbulent Flows," Comp. Meth. Appl. Mech. Eng., 3, 269 (1974).

Ref. 2-16. W. M. Rohsenow and H. Choi, Heat, Mass, and Momentum Transfer (Prentice-Hall, Englewood Cliffs, New Jersey, 1961).

Ref. 2-17. H. Schlicting, Boundary-Layer Theory (McGraw-Hill, New York, 1968).

Ref. 2-18. R. S. Brodkey, The Phenomena of Fluid Motions (Addison Wesley, Boston, 1969).

Ref. 2-19. David C. Wilcox, Turbulence Modeling for CFD (Griffin Printing, Glendale, California, 1944).

Ref. 2-20. T. Cebeci and A. M. O. Smith, Analysis of Turbulent Boundary Layers(Academic Press, New York, 1974).

Ref. 2-21. W. Rodi, Turbulence Models and Their Application in Hydraulics (International Association for Hydraulic Research, Delft, The Netherlands, 1980).

Ref. 2-22. P. Bradshaw, T. Cebeci, and J. H. Whitelaw, Engineering Calculation Methods in Turbulent Flow (Academic Press, New York, 1981).

Ref. 2-23. J. R. Travis, "A Heat, Mass, and Momentum Transport Model for Hydrogen Diffusion Flames in Nuclear Reactor Containments," Nuclear Engr. Des., 101, 149 (1987).

Ref. 2-24. F. Tamanini, E. A. Ural, and J. L. Chaffee, "Hydrogen Combustion Experiments in a 1/4-scale Model of a MARK III Nuclear Reactor Containment," Factory Mutual Corporation Interim report Y101-1 (May 1987).

Ref. 2-25. P. Royl, "Hydrogen Distribution and Combustion - Validation and Application of the GASFLOW-MPI Code," SMiRT-13 Post Conference Seminar No. 4 'Containment Facilities', Brazilia, Brazil (August 21-22, 1995).

Ref. 2-26. L. Valencia, "Brandschutzversuche im Containment HDR-Versuchsgruppe BRA Versuche T52.1/T52.2" (Fire Protection Tests in the Containment HDR-Test Group BRA Tests T52.1/T52.2), Kernforschungszentrum Karlsruhe report PHDR-Arbeitsbericht Nr. 5.075/86 (February 1987).

Ref. 2-27. E. S. Oran, T. R. Young, and J. P. Boris, Seventeenth Symposium (International) on Combustion, p. 43 (The Combustion Institute, 1978).

Ref. 2-28. T. R. Young, "CHEMEQ - A Subroutine for Solving Stiff Ordinary Differential Equations," Naval Research Laboratory Memorandum Report 4091 (1979).

Ref. 2-29. B. F. Magnussen and B. H. Hjertager, "On Mathematical Modelling of Turbulent Combustion with Special Emphasis on Soot Formation and Combustion," Sixteenth Symposium (International) on Combustion, pp. 719-729 (The Combustion Institute, Pittsburgh, 1976).

Ref. 2-30. B. H. Hjertager, Numerical Simulation of Turbulent Flame and Pressure Development in Gas Explosions in Fuel-Air Explosions, SM Study No. 16, pp. 407-426 (University of Waterloo Press, Ontario, Canada, 1982).

Ref. 2-31. T. Kanzleiter, "Versuche zur Wirksamkeit von Wasserstroff-Gegenmassnahmen in einer Mehrraum-Containment-Geometrie, Band 1 und Band 2," Battelle Frankfurt abschlussbericht BIeV-R67.036-01 und 02 (November 1991).

Ref. 2-32. U. Behrens, G. Ahrens, R.E. Schmitt, and L. Wandzilak, "Experimentelle Untersuchungen zum Verhalten des vom NIS entwickelten Katalysator-Moduls im 1:1-Mabstab bei verschiedenen Systemzuständen im Modell-Containment," Battelle Frankfurt report (June 1991).

Ref. 2-33. K. Fischer, "Qualification of a Passive Catalytic Module for Hydrogen Mitigation," Nuclear Technology, 112, (Oct. 1995).

Ref. 2-34. W. Breitung, "Protocoll der AG-Sitzung H_2-Verbrennung, EVU/KfK/Siemens-Vertrag," Projekt Nukleare Sicherheitsforschung Protokoll, (Nov. 1994).

Ref. 2-35. S. G. Markandeya and A. K. Chakraborty, "Modelling of Catalytic Recombiners for Removal of Hydrogen During Severe Accidents," Proceedings of the SMiRT-12 Conference, (Elsevier Science Publishers, 1993).

Ref. 2-36. The Engineering Division, Flow of Fluids through Valves, Fittings, and Pipe, Technical Paper No. 410 (Crane Company, New York, NY, 1976).

Ref. 2-37. B. Liu, D. Mather, and R. D. Reitz. Modeling the Effects of Drop Drag and Breakup on Fuel Sprays. SAE Technical Paper 930072, SAE, 1993.

Ref. 2-38. R. S. Hotchkiss and C. W. Hirt, "Particulate Transport in Highly Distorted Three-Dimensional Flow Fields," Proceedings of the 1972 Summer Computer Simulation Conference Vol. 2, pp. 1037-1041 (AAIA, San Diego, CA, 1972).

Ref. 2-39. William C. Hinds, Aerosol Technology (John Wiley and Sons, New York, 1982).

Ref. 2-40. Barton Dahneke, "The Capture of Aerosol Particles by Surfaces," J. Colloid Interface Sci., 37, 342 (1971).

Ref. 2-41. Barton Dahneke, "The Influence of Flattening on the Adhesion of Particles," J.Colloid Interface Sci., 40, 1 (1972).

Ref. 2-42. R. S. Bradley, Phil. Mag., 13, 853 (1932).

Ref. 2-43. H. C. Hamaker, "The London-Van der Waals Attraction Between Spherical Particles," Physica (Ulrecht), 4, 1058 (1937).

Ref. 2-44. Kyaw Tha Paw U, "The Rebound of Particles from Natural Surfaces," J. Colloid Interface Sci., 93, 442 (1983).

Ref. 2-45. Barton Dahneke, "Measurements of Bouncing of Small Latex Spheres," J. Colloid Interface Sci., 45, 584 (1973).

Ref. 2-46. Barton Dahneke, "Further Measurements of the Bouncing of Small Latex Spheres," J. Colloid Interface Sci., 51, 58 (1975).

Ref. 2-47. John S. Halow, "Incipient Rolling, Sliding and Suspension of Particles in Horizontal and Inclined Turbulent Flow," Chem. Engr. Sci., 28, 1 (1973).

Ref. 2-48. P. G. Saffman, "The Lift on a Small Sphere in a Slow Shear Flow," J. Fluid Mech., 22, 385 (1965).

Ref. 2-49. J. D. Iversen, J. B. Pollack, R. Greeley and B. R. White, "Saltation Threshold on Mars: The Effect of Interparticle Force, Surface Roughness, and Low Atmospheric Density," ICARUS, 29, 381 (1976).

Ref. 2-50 F.R. Menter, Zonal Two Equation k-w Turbulence Models for Aerodynamic Flows, AIAA Paper #93-2906, 24th Fluid Dynamics Conference, July 1993

Ref. 2-51 F.R. Menter, Two-Equation Eddy-Viscosity Turbulence Models for Engineering Applications, AIAA Journal, vol. 32, no. 8, pp. 1598-1605, 1994

3 Computational Model

The computational model and solution algorithm for solving the multidimensional, time-dependent fluid-flow equations follows the ICE'd-ALE methodology first introduced by Hirt et al. (Ref. 3-1) and used later in other computational fluid dynamics efforts at Los Alamos (Ref. 3-1 through Ref. 3-7).

The computational domain is defined by regular, three-dimensional arrays of regular parallelepiped cells in three-dimensional computing domains, and one-dimensional arrays of either square cross-sections or cylindrical cross-sections in the piping or duct networks. First, each coordinate axis is divided into intervals that define cell faces in that direction. The location of each interval is identified by its physical mesh coordinate (*x, y,* or *z* in cartesian geometry or *r*, θ, or *z* in cylindrical geometry) and a corresponding logical coordinate (*i, j,* or *k*) called the mesh index. The domain is divided up in the *x*-direction by planes passing through the *x*-direction mesh coordinates and normal to the *x*-axis. Similarly, sets of planes normal to the other directions divide up space in the *y* -and *z*- directions. The intersections of the three families of planes define a three-dimensional arrangement of cells ("finite volumes"). Figure 3-1 shows a typical cell, along with conventions for identifying faces and vertices.

Values of the scalar variables ρ, *I*, and *p* are computed at cell centers, and the face-normal component of velocity is computed at cell faces or edges. Initial values for each variable and appropriate boundary conditions are set at all locations. The continuous integral equations of motion described in Section 0 are approximated by finite-volume expressions (discrete algebraic equations) on the computational mesh. Thus, the dynamic state in the problem domain can be approximated by integrating the finite-volume equations in space and time.

In the standard ALE method (Ref. 3-1), both fluid and grid ($\mathbf{u_g}$) velocities are located at cell vertices. By specifying $\mathbf{u_g}$ to be different from $\mathbf{u}$, the shape and spatial distribution of the mesh may be changed to model a problem with a deformable boundary. Mass, momentum, and energy are exchanged between cells by averaging vertex velocities to produce a cell-face fluxing velocity. Because GASFLOW-MPI is designed to compute flows with fixed geometries, this general mesh motion feature is not needed; therefore, we locate fluid velocities directly on cell faces. Other differences between the standard ALE method and the ALE method implemented in GASFLOW-MPI are given in Table 3-1.

Mesh cells serve as control volumes for cell-centered variables and thus serve as the finite volumes for solving the mass and energy equations. Because velocity components are located at cell faces, a different treatment is needed for the momentum equations. A momentum control volume, V_m, is defined as half of each of the two cells sharing a common face (Figure 3-2). In this sense, we say that momentum control volumes are "face-centered," although this is strictly true only if both cells are the same size. A momentum control volume for the east face of cell (*i, j, k*) is $V_m = (V_{i,j,k} + V_{i+1,j,k})/2$. A similar definition is made for the momentum control volumes that straddle the north and top faces.

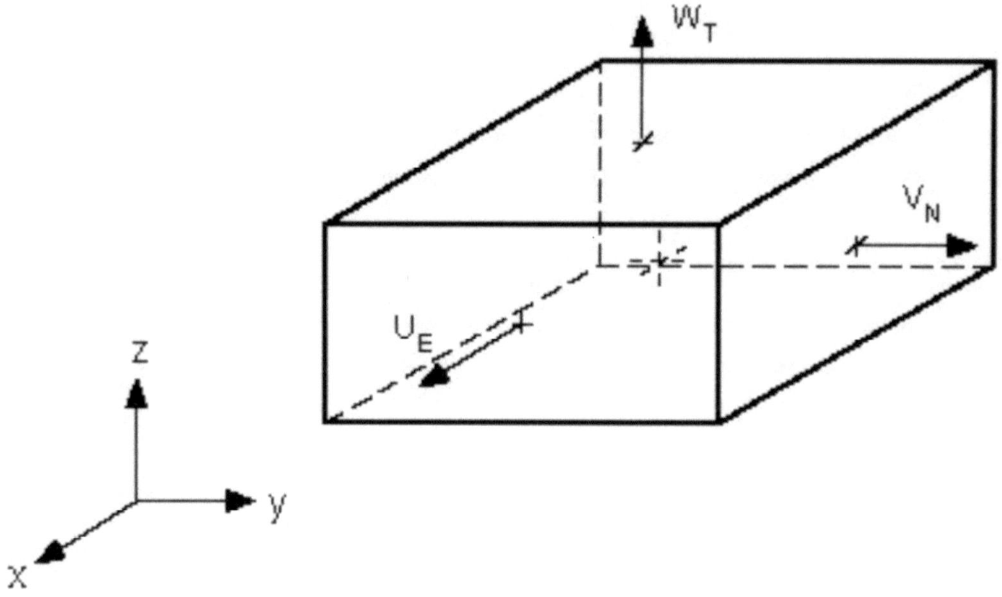

Figure 3-1 GASFLOW-MPI three-dimensional computational cell. Velocity components are located at face centers: [W] west (-x), [E] east (+x), [S] south (-y), [N] north (+y), [B] bottom (-z), and [T] top (+z); all other variables, the scalar variables, are located at the cell center.

Table 3-1 Comparison of standard ALE and GASFLOW-MPI ALE features

Standard ICE'd ALE	GASFLOW-MPI linearized ICE'd ALE
Cell-centered ρ, I, p	Cell-centered ρ, I, p
Vertex-centered u, momentum	Face-centered u, momentum
Total energy (E)	Internal energy (I)
Pressure work is done only in the implicit Lagrangian phase.	Pressure work is done in both the explicit and implicit Lagrangian phases.
Control volumes are (1) computational cells and (2) volumes centered on vertices.	Control volumes are (1) computational cells and (2) volumes centered on cell faces.
Vertex motion algorithm	Vertices stay fixed.
Arbitrary hexahedral cells	Parallelepiped cells
Algorithm for cell volumes	Volume calculations are straightforward.
Face fluxing in rezone phase (face-fluxing u must be computed from vertex u)	*x*-, *y*-, and *z*-face fluxing in rezone phase
	Linearized volume treatment in implicit Lagrangian phase

Because we are interested only in the Eulerian solution of the flow equations, a full continuous rezone always will be applied (see Section 3.4, Phase C: Rezone Phase), with the Lagrangian phase being only an intermediate step toward the full solution.

Each fluid dynamics time step is broken into three phases as described below and is followed by turbulent transport calculations.

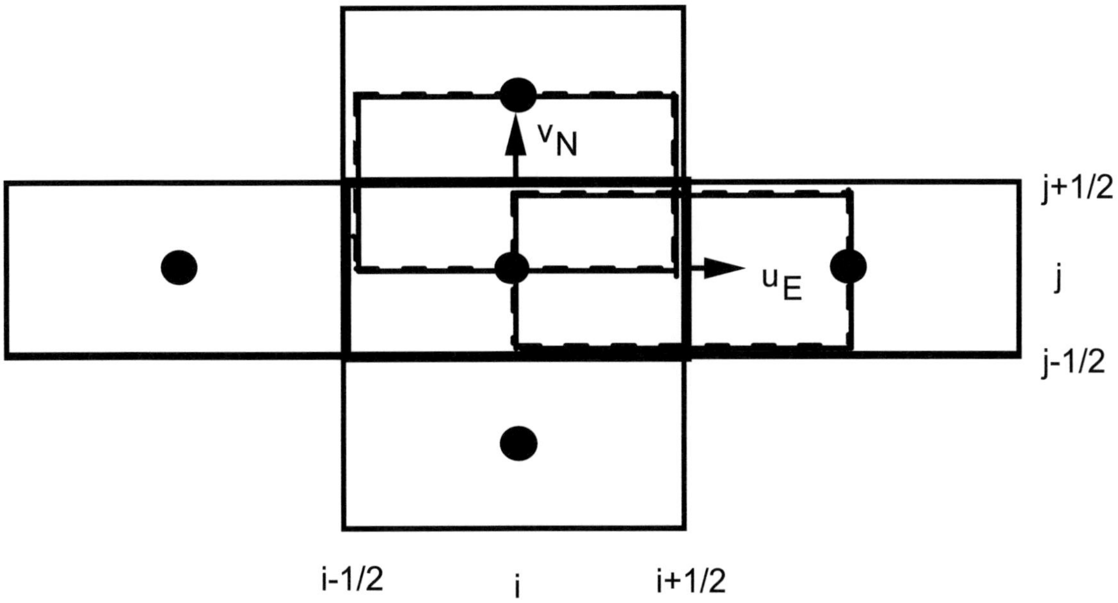

Figure 3-2. A two-dimensional mesh slice showing the u and v staggered momentum control volumes associated with cell (i, j, k).

3.1 Beginning of Time Cycle Initialization

GASFLOW-MPI allows the user to compute molecular transport properties by a number of methods. The first and simplest is a "nonmechanistic" calculation using constant default or input variables such as kinematic viscosity, thermal conductivity, binary mass diffusion coefficient, and the non-dimensional Prandtl and Schmidt numbers. These option are discussed in detail in the GASFLOW-MPI User's Manual (NUREG/CR-6570, Vol. 2). Based upon the gas composition of each finite control volume, the mixture transport properties are updated by the procedures outlined in Section 2.5.1.2 (Transport Properties).

If the user has specified a turbulence model, then the molecular transport coefficients are updated to the apparent or turbulent values using the methodology described in Section 2.7.4 (Turbulence Effects on the Transport Coefficients).

3.2 Phase A: Explicit Lagrangian Phase for the Multidimensional Finite Control Volumes

In this phase, the densities, velocities, and specific internal energy fields are updated by the effects of all chemical and physical processes. These effects include combustion, catalytic recombination, heat transfer, phase change between liquid film on structural surfaces and within the fluid mixture, body forces, and turbulence effects.

3.2.1 Volume Equation

The change in each computational cell volume associated with the scalar variable, V_s, is calculated from the discrete approximation to Equation (2-5) with $\Phi = 1$ and application of the divergence theorem :

$$\frac{V^A - V^n}{\Delta t} = \sum_f (uA)_f^n \Delta S_f \quad . \tag{3-1}$$

Superscript A denotes the Lagrangian Phase A, and n denotes the beginning of computation cycle time-level, while subscript f denotes the cell faces of the appropriate control volume.

3.2.2 Momentum Equations

The components of the velocity field then can be found from the discrete approximation of Equation (2-13):

$$\frac{\rho_m^A \mathbf{u}^A V_m^A - \rho_m^n \mathbf{u}^n V_m^n}{\Delta t} = -\sum_f (p)_f^n \mathbf{S}_f + \rho_m^n \mathbf{g} V_m^n - \sum_f (\tau A)_f^n \Delta S_f - \sum_f (\mathbf{D}_d A)_f^A \Delta S_f \quad . \tag{3-2}$$

In addition to the resistance to flow through reduced flow areas, areas that are smaller than the computational cell-face area, the structural drag vector, $\mathbf{D}_d$, can represent the resistance of internal structures such as pipes, I-beams, catwalks, and such configurations that are impossible to resolve on any practical mesh. These internal structures play an important role as heat sinks and, to a lesser degree, as momentum sinks. However, we do attempt to model their momentum effects by the expression

$$\mathbf{D}_d = \frac{1}{2}\mathbf{C}_D (\rho_m \mathbf{u})^A \left|\mathbf{u}^n\right| \quad , \tag{3-3}$$

where the subscript m refers to the staggered momentum control volumes associated with the velocity field. Note that fractional A is dependent on the orientation of the structures. For example, there is little resistance to flows parallel to the gratings of the catwalks but quite a different resistance to flows normal to the gratings.

The drag coefficient, C_D, for reduced area flows is computed from an orifice correlation,

$$\mathbf{C}_D = \left[1 + 0.707 \cdot (1 - \mathbf{A})^{\frac{1}{2}} - \mathbf{A}\right]^2 \quad , \tag{3-4}$$

or can be determined empirically or selected from tables of drag coefficients of common shapes. Equation (3-2) with Equation (3-3) can be solved for the advanced time-level "A" momenta as

$$\rho_m^A \mathbf{u}^A = \frac{\rho_m^n \mathbf{u}^n V_m^n - \Delta t \left[\sum_f (p)_f^n \mathbf{S}_f - \rho_m^n \mathbf{g} V_m^n + \sum_f (\tau A)_f^n \Delta S_f \right]}{V_m^A + \Delta t \mathbf{A} \frac{1}{2} \mathbf{C}_D \left| \mathbf{u}^n \right| S_f} , \tag{3-5}$$

which provides a computationally fast, local implicitness that provides additional robustness to Phase A and the GASFLOW-MPI algorithm. When the mixture density field is updated to the same time-level, as will be shown below, the velocity field can be determined.

3.2.3 Mixture and Species Mass Equations

The mass change for each species because of combustion, phase change, and inter-species diffusion is calculated as follows:

$$\frac{\rho_\alpha^A V^A - \rho_\alpha^n V^n}{\Delta t} = -\sum_f (JA)_f^n \Delta S_f + S_{\rho,\alpha,recombination}^A V^n +$$

$$S_{\rho,\alpha,combustion}^A V^n + S_{\rho,h_2o,mixture}^A V^n + S_{\rho,h_2o,condensation/vaporization}^A V^n + S_{\rho,h_2ol,rainout}^A V^n \quad . \tag{3-6}$$

When Equations (3-2) is summed over all species, we derive the Phase A mixture mass equation:

$$\frac{\rho^A V^A - \rho^n V^n}{\Delta t} = S_{\rho,h_2o,condensation/vaporization}^A V^n + S_{\rho,h_2ol,rainout}^A V^n \quad . \tag{3-7}$$

The "A" time-level on the various source and sink terms will become evident as those models are developed in the sections below.

3.2.4 Internal Energy Equation

The change in the total internal energy can be written from Equation (2-14) as

$$\frac{\rho^A I^A V^A - \rho^n I^n V^n}{\Delta t} = -p^n \sum_f (\mathbf{u} \cdot \mathbf{A})_f^n \Delta S_f + S_{I,recombination}^A V^n +$$

$$\Re_{h_2o} T^A \left(S_{\rho,h_2o,condensation/vaporization}^A + S_{\rho,h_2o,mixture}^A \right) V^n - \sum_f (\mathbf{q} \cdot \mathbf{A})_f^n \Delta S_f +$$

$$S_{I,convection}^A V^n + S_{I,combustion}^A V^n + S_{I,condensation/vaporization}^A V^n + S_{I,rainout}^A V^n \quad . \tag{3-8}$$

All source terms will be derived and discussed in subsequent sections.

3.2.5 Gas Mixture Temperature

The gas-mixture temperature, T^A, is computed by inverting a polynomial of up to the 4th degree, which is similar to Equation (2-18) but cast in a slightly different form

$$A_0 + T^A A_1 + \left(T^A\right)^2 A_2 + \left(T^A\right)^3 A_3 + \left(T^A\right)^4 A_4 = 0 \tag{3-9}$$

such that the polynomial coefficients contain all the know quantities from Equations (3-6) and (3-8). These coefficients will become evident in later developments. The user specifies in the input the degree of the polynomial, from linear to quartic. Our development here will assume the quartic since it contains all the ingredients of the lesser degree polynomials.

3.2.6 Equation of State

Finally, the updated pressure is determined from the equation of state [Equation (2-25)] as follows:

$$p^A = \frac{T^A \sum\limits_{\alpha \neq h_2 ol} R_\alpha \rho_\alpha^A}{\theta_{h_2 ol}^n} , \tag{3-10}$$

where the liquid volume fraction is

$$\theta_{h_2 ol}^n = 1 - \frac{\rho_{h_2 ol}^n}{\rho_{h_2 ol}^{thermo}} . \tag{3-11}$$

3.2.7 Structural Heat Conduction

A heat conducting solid of depth x_D is discretized by N heat conducting elements (1 _ j _ N). Material properties (ρ, c_p, and κ) are independent of temperature in these derivations but thermal conductivity can be different in each element (composite layer). Nodes, also numbered 1 through N, are located on the positive side of the conducting elements. Node zero denotes the surface node on the negative side; node N denotes the surface node on the positive side as shown in Figure 3-3.

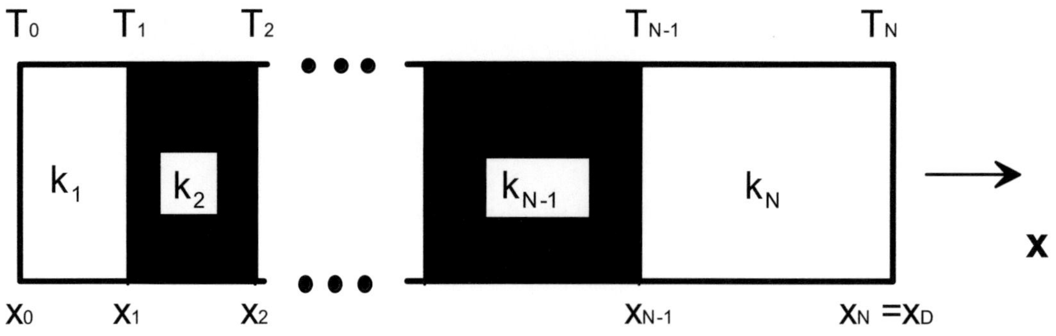

Figure 3-3 Geometry for one-dimensional structural heat conduction

For purposes of problem definition, heat conducting structures are divided into three categories: slabs, sinks, and walls. *Slabs* and *sinks* have only one element in contact with fluid (element No. 1). More accurate spatial resolution of temperatures is generally wanted near the fluid/structure interface; hence, the smallest element is normally element 1 with increasing element sizes away from the interface. *Walls* have both surfaces in contact with a fluid. Comparable discretization of each structural type is shown in Figure 3-4 and Figure 3-5.

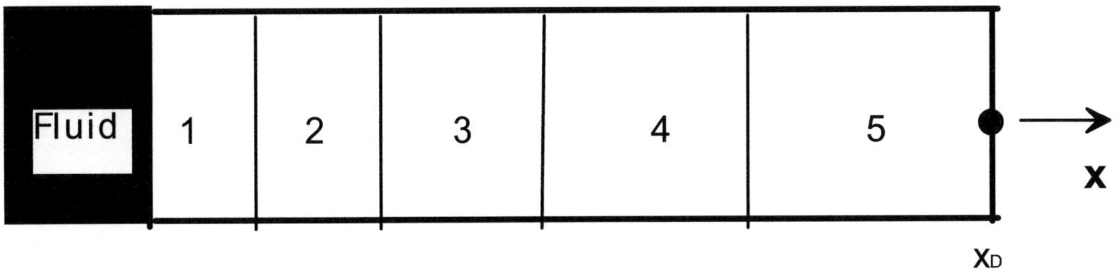

Figure 3-4 Slab and sink discretization

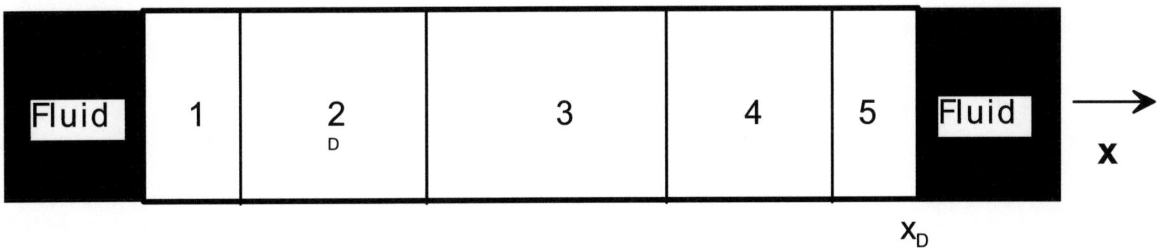

Figure 3-5 Wall discretization

New time-level temperatures at all but the boundary nodes are computed from an implicit finite-difference approximation to Equation (2-52). Details are presented in the following equation:

$$\frac{V\overline{\rho c_p}}{\Delta t}\left(T_j^{n+1}-T_j^n\right) = A\left[\begin{array}{l}\left((1\text{-}\Theta)\left(T_{j+1}^n-T_j^n\right)+\Theta\left(T_{j+1}^{n+1}-T_j^{n+1}\right)\right)k_{j+1}/\Delta x_e \\ -\left((1\text{-}\Theta)\left(T_j^n-T_{j-1}^n\right)+\Theta\left(T_j^{n+1}-T_{j-1}^{n+1}\right)\right)k_j/\Delta x_w\end{array}\right] \quad . \tag{3-12}$$

Three methods are provided for the evaluation of this equation: (1) the temporal fully implicit first-order backward Euler method ($\Theta = 1$), (2) the temporal unconditionally stable second-order Crank-Nicholson method ($\Theta = \frac{1}{2}$), and (3) the temperal fully explicit first-order backward Euler method ($\Theta = 0$). The explicit method is not recommended for containment analyses in which many temporal and spatial scales are encountered.

Different boundary conditions, including specified temperature, specified heat flux, as well as an adiabatic surface, are provided in the solution of Equation (3-12).

Equation (3-12) and two boundary-condition-dependent equations for nodes 0 and *N* are a linear system of *N* equations for the unknown temperatures T_i^{n+1} that are found from a tridiagonal matrix solution algorithm.

3.2.8 Mass and Energy Transfer

3.2.8.1 Structural Convective Heat Transfer

Whenever the temperatures of structural surfaces and the adjacent gas differ there will be heat exchange by convection. In the case when neither water component is included in the calculation, all species' densities remain unchanged during this calculation. The resulting change in energy of the gas appears as source terms in the Phase A energy Equation (3-8) with Equation (2-27),

$$\frac{\rho^A I^A V^A - \rho^n I^n V^n}{\Delta t} = S_{I,convection}^A V^n = \sum_s h_s A_s \left(T_s^n - T^A\right) \quad , \tag{3-13}$$

and as a heat flux boundary condition in the structural heat conduction Equation (3-12). The uncorrected (there is no water component in this case) heat transfer coefficient is computed from Equation (2-29). Making use of Equation (2-18) gives a polynomial of up to degree 4,

$$\left[\sum_\alpha \left(\rho_\alpha^A a_\alpha\right) - \rho^n I^n V^n - \frac{\Delta t}{V^A}\sum_s h_s A_s T_s^n\right] + T^A\left[\sum_\alpha \left(\rho_\alpha^A b_\alpha\right) + \frac{\Delta t}{V^A}\sum_s h_s A_s\right] +$$

$$\left(T^A\right)^2 \sum_\alpha \left(\rho_\alpha^A c_\alpha\right) + \left(T^A\right)^3 \sum_\alpha \left(\rho_\alpha^A d_\alpha\right) + \left(T^A\right)^4 \sum_\alpha \left(\rho_\alpha^A e_\alpha\right) = 0 \quad , \tag{3-14}$$

which must be inverted to find the new mixture temperature. This equation demonstrates the combination of terms that make up the coefficients presented in Equation (3-9). Once the mixture temperature is found the source term in the energy equation can be computed from

$$S_{I,convection}^A V^n = \sum_s h_s A_s \left(T_s^n - T^A\right) \quad , \tag{3-15}$$

and, therefore, the energy convected to or from any given surface is

$$q_{s,total} = q_{s,convection} = h_s A_s \left(T^A - T_s^n\right) \quad . \tag{3-16}$$

3.2.8.2 Structural Convective Heat and Mass Transfer

When the water vapor component is specified with user input, phase change on structural surfaces may occur. In this case, the water vapor species mass Equation (3-6) with Equation (2-42) is

$$\frac{\rho_{h_2o}^A V^A - \rho_{h_2o}^n V^n}{\Delta t} = S_{\rho,h_2o,condensation/vaporization}^A V^n =$$

$$-\sum_{s} \max\left[-\frac{\delta_s^n A_s \rho_{h_2ol}^{thermo}}{2\Delta t}, h_d^* A_s \left(\rho_{h_2o}^A - \rho_{s,saturation}\right)\right] \quad , \tag{3-17}$$

and the energy Equation (3-8) with Equations (2-27) and (2-44) becomes

$$\frac{\rho^A I^A V^A - \rho^n I^n V^n}{\Delta t} = \Re_{h_2o} T^A S^A_{\rho,h_2o,condensation/vaporization} V^n + S^A_{I,convection} V^n +$$

$$S^A_{I,condensation/vaporization} V^n =$$

$$\Re_{h_2o} T^A S^A_{\rho,h_2o,condensation/vaporization} V^n + \sum_{s} h_s A_s \left(T_s^n - T^A\right) +$$

$$-\sum_{s} \max\left[\begin{array}{l} -\dfrac{\delta_s^n A_s \rho_{h_2ol}^{thermo}}{2\Delta t} I_{h_2o}\left(T_s\right), \\ h_d^* A_s \left(\rho_{h_2o}^A - \rho_{s,saturation}\right) I_{h_2o}\left(T_s\right), \\ h_d^* A_s \left(\rho_{h_2o}^A - \rho_{s,saturation}\right) I_{h_2o}\left(T^A\right) \end{array}\right] \quad . \tag{3-18}$$

The enhanced or corrected heat and mass transfer coefficients are computed as described in Sections 2.6.1 and 2.6.2, and the surface saturation density is evaluated by the formulation given in Section 2.6.2.

The solution procedure for Equations (3-17) and (3-18) requires several passes through the mesh. For example, the term on the right-hand side of Equation (3-17) checks to see if surface dryout occurs (the liquid film is depleted); this must be determined without knowing the advanced time-level water vapor density. One could employ a iterative scheme, but we have found a much faster solution procedure: we make a predictive pass through the mesh and then, based upon the predictive information, a correction pass is performed.

First, a fully local implicit evaluation of Equation (3-17) is executed without regard to surface dryout to find a predicted density, $\rho_{h_2o}^p$, by reducing Equation (3-17) to

$$\frac{\rho_{h_2o}^p V^A - \rho_{h_2o}^n V^n}{\Delta t} = -\sum_{s} h_d^* A_s \left(\rho_{h_2o}^p - \rho_{s,saturation}\right) \quad . \tag{3-19}$$

The solution for the predictor water vapor density, $\rho_{h_2o}^p$, is

$$\rho_{h_2o}^p = \frac{\rho_{h_2o}^n V^n + \Delta t \sum_{s} h_d^* A_s \rho_{s,saturation}}{V^A + \Delta t \sum_{s} h_d^* A_s} \quad . \tag{3-20}$$

The corrective step makes use of the predicted density, $\rho_{h_2o}^{p}$, in the following way:

$$\frac{\rho_{h_2o}^{A} V^{A} - \rho_{h_2o}^{n} V^{n}}{\Delta t} =$$

$$- \sum_{s}^{evaporation} \left[1 - H\left(\rho_{h_2o}^{p} - \rho_{s,saturation} \right) \right] \cdot \max \begin{bmatrix} -\dfrac{\delta_s^n A_s \rho_{h_2ol}^{thermo}}{2\Delta t}, \\ h_d^* A_s \left(\rho_{h_2o}^{p} - \rho_{s,saturation} \right) \end{bmatrix}$$

$$- \sum_{s}^{condensation} H\left(\rho_{h_2o}^{p} - \rho_{s,saturation} \right) \cdot \max \begin{bmatrix} 0, \\ h_d^* A_s \left(\rho_{h_2o}^{A} - \rho_{s,saturation} \right) \end{bmatrix} , \tag{3-21}$$

where $H\left(\rho_{h_2o}^{p} - \rho_{s,saturation} \right)$ is the usual Heaviside function

$$H\left(\rho_{h_2o}^{p} - \rho_{s,saturation} \right) = \begin{cases} 0; \rho_{h_2o}^{p} - \rho_{s,saturation} < 0 \\ 1; \rho_{h_2o}^{p} - \rho_{s,saturation} > 0 \end{cases} . \tag{3-22}$$

Note that the first term on the right-hand side of the equation acts as a source of water vapor because evaporation of the liquid film is occurring, whereas the second term on the right-hand side is a loss of water vapor mass due to condensation of water vapor in the mixture. Since this second term could result in negative water vapor density, it can be treated implicitly with the water vapor source term being treated as a constant. The corrected water vapor density is then computed as follows:

$$\rho_{h_2o}^{A} = \frac{\rho_{h_2o}^{n} V^{n} - \Delta t \left\{ \begin{matrix} \displaystyle\sum_{s}^{evaporation} \left[1 - H\left(\rho_{h_2o}^{p} - \rho_{s,saturation} \right) \right] \cdot \max \begin{bmatrix} -\dfrac{\delta_s^n A_s \rho_{h_2ol}^{thermo}}{2\Delta t}, \\ h_d^* A_s \left(\rho_{h_2o}^{p} - \rho_{s,saturation} \right) \end{bmatrix} \\ \displaystyle - \sum_{s}^{condensation} H\left(\rho_{h_2o}^{p} - \rho_{s,saturation} \right) \cdot h_d^* A_s \rho_{s,saturation} \end{matrix} \right\}}{V^{A} + \Delta t \displaystyle\sum_{s}^{condensastion} H\left(\rho_{h_2o}^{p} - \rho_{s,saturation} \right) \cdot h_d^* A_s} . \tag{3-23}$$

The actual source term is then computed as follows:

$$S_{\rho,h_2o,condensation/vaporization}^{A} V^{n} =$$

$$- \sum_{s}^{evaporation} \left[1 - H\left(\rho_{h_2o}^{p} - \rho_{s,saturation} \right) \right] \cdot \max \begin{bmatrix} -\dfrac{\delta_s^n A_s \rho_{h_2ol}^{thermo}}{2\Delta t}, \\ h_d^* A_s \left(\rho_{h_2o}^{p} - \rho_{s,saturation} \right) \end{bmatrix}$$

$$- \sum_{s}^{condensation} H\left(\rho_{h_2o}^{p} - \rho_{s,saturation} \right) \cdot \max \begin{bmatrix} 0, \\ h_d^* A_s \left(\rho_{h_2o}^{A} - \rho_{s,saturation} \right) \end{bmatrix} , \tag{3-24}$$

where the mass transfer for each surface is

$$\dot{m}_s = \left[1 - H\left(\rho^p_{h_2o} - \rho_{s,saturation}\right)\right] \cdot \max\begin{bmatrix} -\dfrac{\delta^n_s A_s \rho^{thermo}_{h_2ol}}{2\Delta t}, \\ h^*_d A_s \left(\rho^p_{h_2o} - \rho_{s,saturation}\right) \end{bmatrix}$$

$$+H\left(\rho^p_{h_2o} - \rho_{s,saturation}\right) \cdot \max\begin{bmatrix} 0, \\ h^*_d A_s \left(\rho^A_{h_2o} - \rho_{s,saturation}\right) \end{bmatrix} \quad . \tag{3-25}$$

The liquid film thickness, δ^A_s, for each surface is finally updated using the mass transfer computed by Equation (3-25):

$$\delta^A_s = \delta^n_s + \frac{\dot{m}_s \Delta t}{A_s \rho^{thermo}_{h_2ol}} \quad . \tag{3-26}$$

The energy Equation (3-18) can be evaluated to determine the mixture temperature by the following function with respect to T:

$$\left\{\begin{matrix} \sum\limits_\alpha \left(\rho^A_\alpha a_\alpha\right) - \dfrac{1}{V^A}\rho^n I^n V^n - \dfrac{\Delta t}{V^A}\sum\limits_s h_s A_s T^n_s + \\ \dfrac{\Delta t}{V^A}\sum\limits_s \left[1 - H\left(\rho^p_{h_2o} - \rho_{s,saturation}\right)\right] \cdot \max\begin{bmatrix} -\dfrac{\delta A_s \rho^{thermo}_{h_2ol}}{2\Delta t} I_{h_2o}\left(T_s\right), \\ h^*_d A_s \left(\rho^p_{h_2o} - \rho_{s,saturation}\right) I_{h_2o}\left(T_s\right) \end{bmatrix} + \\ \dfrac{\Delta t}{V^A}\sum\limits_s H\left(\rho^p_{h_2o} - \rho_{s,saturation}\right) \cdot \max\begin{bmatrix} 0, \\ a_{h_2o} h^*_d A_s \left(\rho^A_{h_2o} - \rho_{s,saturation}\right) \end{bmatrix} \end{matrix}\right\} +$$

$$T^A \left\{\begin{matrix} \sum\limits_\alpha \left(\rho^A_\alpha b_\alpha\right) - \dfrac{\Delta t}{V^A} R_{h_2o} S^A_{\rho,h_2o,condensation/vaporization} V^n + \dfrac{\Delta t}{V^A}\sum\limits_s h_s A_s + \\ \dfrac{\Delta t}{V^A}\sum\limits_s H\left(\rho^p_{h_2o} - \rho_{s,saturation}\right) \cdot \max\begin{bmatrix} 0, \\ b_{h_2o} h^*_d A_s \left(\rho^A_{h_2o} - \rho_{s,saturation}\right) \end{bmatrix} \end{matrix}\right\} +$$

$$\left(T^A\right)^2 \left\{\begin{matrix} \sum\limits_\alpha \left(\rho^A_\alpha c_\alpha\right) + \\ \dfrac{\Delta t}{V^A}\sum\limits_s H\left(\rho^p_{h_2o} - \rho_{s,saturation}\right) \cdot \max\begin{bmatrix} 0, \\ c_{h_2o} h^*_d A_s \left(\rho^A_{h_2o} - \rho_{s,saturation}\right) \end{bmatrix} \end{matrix}\right\} +$$

$$\left(T^A\right)^3 \left\{ \begin{array}{l} \sum_{\alpha}\left(\rho_{\alpha}^{A} d_{\alpha}\right)+ \\ \frac{\Delta t}{V^A}\sum_{s} H\left(\rho_{h_2o}^{p}-\rho_{s,saturation}\right)\cdot \max\left[\begin{array}{l} 0, \\ d_{h_2o} h_d^* A_s\left(\rho_{h_2o}^{A}-\rho_{s,saturation}\right) \end{array}\right] \end{array} \right\}+$$

$$\left(T^A\right)^4 \left\{ \begin{array}{l} \sum_{\alpha}\left(\rho_{\alpha}^{A} e_{\alpha}\right)+ \\ \frac{\Delta t}{V^A}\sum_{s} H\left(\rho_{h_2o}^{p}-\rho_{s,saturation}\right)\cdot \max\left[\begin{array}{l} 0, \\ e_{h_2o} h_d^* A_s\left(\rho_{h_2o}^{A}-\rho_{s,saturation}\right) \end{array}\right] \end{array} \right\}=0 \quad . \tag{3-27}$$

Once the mixture density and temperature are found, we can determine each of the energy source terms found in Equation (3-18) as follows:

$$R_{h_2o} T^A S^A_{\rho,h_2o,condensation/vaporization} V^n =$$

$$-R_{h_2o} T^A \left\{ \begin{array}{l} \sum_{s}^{evaporation}\left[1-H\left(\rho_{h_2o}^{p}-\rho_{s,saturation}\right)\right]\cdot \max\left[\begin{array}{l} -\frac{\delta_s^n A_s \rho_{h_2ol}^{thermo}}{2\Delta t}, \\ h_d^* A_s\left(\rho_{h_2o}^{p}-\rho_{s,saturation}\right) \end{array}\right] \\ +\sum_{s}^{condensation} H\left(\rho_{h_2o}^{p}-\rho_{s,saturation}\right)\cdot \max\left[\begin{array}{l} 0, \\ h_d^* A_s\left(\rho_{h_2o}^{A}-\rho_{s,saturation}\right) \end{array}\right] \end{array} \right\} \tag{3-28}$$

$$S^A_{I,convection} V^n = \sum_s h_s A_s \left(T_s^n - T^A\right) \tag{3-29}$$

$$S^A_{I,condensation/vaporization} V^n =$$

$$-\sum_{s}^{evaporation}\left[1-H\left(\rho_{h_2o}^{p}-\rho_{s,saturation}\right)\right]\cdot \max\left[\begin{array}{l} -\frac{\delta_s^n A_s \rho_{h_2ol}^{thermo}}{2\Delta t}, \\ h_d^* A_s\left(\rho_{h_2o}^{p}-\rho_{s,saturation}\right) \end{array}\right]$$

$$-\sum_{s}^{condensation} H\left(\rho_{h_2o}^{p}-\rho_{s,saturation}\right)\cdot \max\left[\begin{array}{l} 0, \\ h_d^* A_s\left(\rho_{h_2o}^{A}-\rho_{s,saturation}\right) \end{array}\right] \quad . \tag{3-30}$$

The total energy, $q_{s,total}$, delivered to a given surface is

$$q_{s,total} = q_{s,convection} + q_{s,condensation/vaporization} =$$

$$h_s A_s \left(T^A - T_s^n\right) + \left[1 - H\left(\rho_{h_2o}^p - \rho_{s,saturation}\right)\right] \left\{ \max\left[\begin{array}{l} -\dfrac{\delta_s^n A_s \rho_{h_2ol}^{thermo}}{2\Delta t}, \\ h_d^* A_s \left(\rho_{h_2o}^p - \rho_{s,saturation}\right) \end{array} \right] \cdot \left[h_{fg} + I_{h_2o}\left(T_s^n\right)\right] \right\}$$

$$+ H\left(\rho_{h_2o}^p - \rho_{s,saturation}\right) \left\{ \max\left[\begin{array}{l} 0, \\ h_d^* A_s \left(\rho_{h_2o}^A - \rho_{s,saturation}\right) \end{array} \right] \cdot \left[h_{fg} + I_{h_2o}\left(T^A\right) - I_{h_2o}\left(T_s^n\right)\right] \right\} . \qquad (3\text{-}31)$$

3.2.8.3 Structural Convective Heat and Mass Transfer with Phase Change in the Fluid Mixture HEM

When the water vapor and liquid components are specified with user input, phase change on structural surfaces may occur plus phase change with the fluid mixture may also occur. In this case, the water vapor species mass Equation (3-6) with Equations (2-42) and (2-45) is

$$\frac{\rho_{h_2o}^A V^A - \rho_{h_2o}^n V^n}{\Delta t} = S_{\rho,h_2o,condensation/vaporization}^A V^n + S_{\rho,h_2o,mixture}^A V^n =$$

$$-\sum_s \max\left[-\frac{\delta_s^n A_s \rho_{h_2ol}^{thermo}}{2\Delta t}, h_d^* A_s \left(\rho_{h_2o}^A - \rho_{s,saturation}\right) \right]$$

$$-V^n C \cdot \max\left[\begin{array}{l} -\rho_{h_2ol}^n / 10, \\ \rho_{h_2o}^A - \rho_{saturation}\left(T, p_{saturation}\right) \end{array} \right] , \qquad (3\text{-}32)$$

Divide by 2 instead of 10 in last term. Explain further:

the water liquid species mass Equation (3-6) is

$$\frac{\rho_{h_2ol}^A V^A - \rho_{h_2ol}^n V^n}{\Delta t} = S_{\rho,h_2ol,fluid}^A V^n = -S_{\rho,h_2o,mixture}^A V^n =$$

$$+V^n C \cdot \max\left[\begin{array}{l} -\rho_{h_2ol}^n / 10, \\ \rho_{h_2o}^A - \rho_{saturation}\left(T, p_{saturation}\right) \end{array} \right] , \qquad (3\text{-}33)$$

and the energy Equation (3-8) with Equations (2-27), (2-44), and (2-49) is

$$\frac{\rho^A I^A V^A - \rho^n I^n V^n}{\Delta t} = \Re_{h_2o} T^A \left(S^A_{\rho,h_2o,condensation/vaporization} + S^A_{\rho,h_2o,mixture} \right) V^n +$$

$$S^A_{I,condensation/vaporization} V^n + S^A_{I,convection} V^n +$$

$$R_{h_2o} T^A \left(S^A_{\rho,h_2o,condensation/vaporization} + S^A_{\rho,h_2o,mixture} \right) V^n + \sum_s h_s A_s \left(T^n_s - T^A \right) +$$

$$-\sum_s \max \begin{bmatrix} -\dfrac{\delta^n_s A_s \rho^{thermo}_{h_2ol}}{2\Delta t} I_{h_2o}(T_s), \\ h^*_d A_s \left(\rho^A_{h_2o} - \rho_{s,saturation} \right) I_{h_2o}(T_s), \\ h^*_d A_s \left(\rho^A_{h_2o} - \rho_{s,saturation} \right) I_{h_2o}(T^A) \end{bmatrix} . \tag{3-34}$$

The enhanced or corrected heat and mass transfer coefficients are computed as described in Sections 2.6.1 and 2.6.2, the surface saturation density is evaluated by the formulation given in Section 2.6.2, and the saturation density in the fluid mixture is computed from Equation (2-48) in Section 2.6.3.

The solution procedure for Equation (3-32) is very similar to the solution procedure shown in the previous section which required several passes through the mesh. For example, the first term on the right-hand side of Equation (3-32) checks to see if surface dryout occurs (the liquid film is depleted), whereas the second term accounts for phase change within the fluid mixture. We again implement a procedure to make a predictive pass through the mesh, and then, based upon the predictive information, a correction pass is performed.

First, a fully local implicit evaluation of Equation (3-32) is executed without regard to surface dryout to find a predicted density, $\rho^p_{h_2o}$, by reducing Equation (3-32) to the following:

$$\frac{\rho^p_{h_2o} V^A - \rho^n_{h_2o} V^n}{\Delta t} =$$

$$-\sum_s h^*_d A_s \left(\rho^p_{h_2o} - \rho_{s,saturation} \right) - V^n C \left[\rho^p_{h_2o} - \rho_{saturation}(T, p_{saturation}) \right] . \tag{3-35}$$

The solution to this equation is

$$\rho^p_{h_2o} = \frac{\rho^n_{h_2o} V^n + \Delta t \left\{ \sum_s h^*_d A_s \rho_{s,saturation} + V^n C \rho_{saturation}(T, p_{saturation}) \right\}}{V^A + \Delta t \left\{ \sum_s h^*_d A_s + V^n C \right\}} . \tag{3-36}$$

The corrective step makes use of the predicted density, $\rho_{h_2o}^{p}$, in the following way:

$$\frac{\rho_{h_2o}^{A}V^{A}-\rho_{h_2o}^{n}V^{n}}{\Delta t}=$$

$$-\sum_{s}^{evaporation}\left[1-H\left(\rho_{h_2o}^{p}-\rho_{s,saturation}\right)\right]\cdot\max\begin{bmatrix}-\dfrac{\delta_s^n A_s\rho_{h_2ol}^{thermo}}{2\Delta t}, \\ h_d^* A_s\left(\rho_{h_2o}^{p}-\rho_{s,saturation}\right)\end{bmatrix}$$

$$-\sum_{s}^{condensation} H\left(\rho_{h_2o}^{p}-\rho_{s,saturation}\right)\cdot\max\begin{bmatrix}0, \\ h_d^* A_s\left(\rho_{h_2o}^{A}-\rho_{s,saturation}\right)\end{bmatrix}$$

$$-V^nC\left\{\begin{matrix}\left[1-H\left(\rho_{h_2o}^{p}-\rho_{sat}(T^n,p_{sat})\right)\right]\cdot\max\begin{bmatrix}-\rho_{h_2ol}^{n}/10, \\ \rho_{h_2o}^{p}-\rho_{sat}(T^n,p_{sat})\end{bmatrix} \\ +H\left(\rho_{h_2o}^{p}-\rho_{sat}(T^n,p_{sat})\right)\cdot\max\begin{bmatrix}0, \\ \rho_{h_2o}^{A}-\rho_{sat}(T^n,p_{sat})\end{bmatrix}\end{matrix}\right\}. \quad (3\text{-}37)$$

Using the same reasoning as in the previous section, we note that the first term on the right-hand side of the equation acts as a source of water vapor because evaporation of the liquid film is occurring, whereas the second term on the right-hand side is a loss of water vapor mass due to condensation of water vapor from the mixture. The third term is the exchange of mass between the liquid and vapor phases, which can be either a source or sink in this equation. Since the second and third terms could result in negative water vapor density, they can be treated implicitly with the water vapor source term being treated as a constant.

Solving Equation (3-35) for $\rho_{h_2o}^{A}$ yields

$$\rho_{h_2o}^{A}=\frac{\rho_{h_2o}^{n}V^{n}-\Delta t\left\{\begin{matrix}V^nC\cdot\left[1-H\left(\rho_{h_2o}^{p}-\rho_{sat}(T^n,p_{sat})\right)\right]\cdot\max\begin{bmatrix}-\rho_{h_2ol}^{n}/10, \\ \rho_{h_2o}^{p}-\rho_{sat}(T^n,p_{sat})\end{bmatrix} \\ +\sum_{s}^{evaporation}\left[1-H\left(\rho_{h_2o}^{p}-\rho_{s,sat}\right)\right]\cdot\max\begin{bmatrix}-\dfrac{\delta_s^n A_s\rho_{h_2ol}^{thermo}}{2\Delta t}, \\ h_d^* A_s\left(\rho_{h_2o}^{p}-\rho_{s,sat}\right)\end{bmatrix} \\ -\sum_{s}^{condensation} H\left(\rho_{h_2o}^{p}-\rho_{s,saturation}\right)\cdot h_d^* A_s\rho_{s,saturation}\end{matrix}\right\}}{V^A+\Delta t\left(V^nC\cdot\left[1-H\left(\rho_{h_2o}^{p}-\rho_{sat}(T^n,p_{sat})\right)\right]+\sum_{s}^{condensastion} H\left(\rho_{h_2o}^{p}-\rho_{s,saturation}\right)\cdot h_d^* A_s\right)}. \quad (3\text{-}38)$$

The actual source terms are then computed as follows:

$$S^A_{\rho,h_2o,condensation/vaporization}V^n =$$

$$-\sum_s^{evaporation}\left[1-H\left(\rho^p_{h_2o}-\rho_{s,saturation}\right)\right]\cdot\max\begin{bmatrix}-\dfrac{\delta^n_s A_s\rho^{thermo}_{h_2ol}}{2\Delta t}, \\ h^*_d A_s\left(\rho^p_{h_2o}-\rho_{s,saturation}\right)\end{bmatrix}$$

$$-\sum_s^{condensation} H\left(\rho^p_{h_2o}-\rho_{s,saturation}\right)\cdot\max\begin{bmatrix}0, \\ h^*_d A_s\left(\rho^A_{h_2o}-\rho_{s,saturation}\right)\end{bmatrix} \quad . \tag{3-39}$$

The mass transfer between phases in the mixture is

$$S^A_{\rho,h_2o,mixture}V^n = -V^n C\begin{Bmatrix}\left[1-H\left(\rho^p_{h_2o}-\rho_{sat}(T^n,p_{sat})\right)\right]\cdot\max\begin{bmatrix}-\rho^n_{h_2ol}/10, \\ \rho^p_{h_2o}-\rho_{sat}(T^n,p_{sat})\end{bmatrix} \\ +H\left(\rho^p_{h_2o}-\rho_{sat}(T^n,p_{sat})\right)\cdot\max\begin{bmatrix}0, \\ \rho^A_{h_2o}-\rho_{sat}(T^n,p_{sat})\end{bmatrix}\end{Bmatrix} \tag{3-40}$$

Just as in the previous section, the actual mass transfer to or from any given surface is

$$\dot{m}_s = \left[1-H\left(\rho^p_{h_2o}-\rho_{s,saturation}\right)\right]\cdot\max\begin{bmatrix}-\dfrac{\delta^n_s A_s\rho^{thermo}_{h_2ol}}{2\Delta t}, \\ h^*_d A_s\left(\rho^p_{h_2o}-\rho_{s,saturation}\right)\end{bmatrix}$$

$$+H\left(\rho^p_{h_2o}-\rho_{s,saturation}\right)\cdot\max\begin{bmatrix}0, \\ h^*_d A_s\left(\rho^A_{h_2o}-\rho_{s,saturation}\right)\end{bmatrix} \quad . \tag{3-41}$$

The liquid film thickness, δ^A_s, for each surface is finally updated using the mass transfer computed by Equation (3-41):

$$\delta^A_s = \delta^n_s + \frac{\dot{m}_s\Delta t}{A_s\rho^{thermo}_{h_2ol}} \quad . \tag{3-42}$$

The energy Equation (3-34) can be evaluated to determine the mixture temperature by inverting the following functional with respect to T:

$$
\left\{
\begin{array}{l}
\sum_{\alpha}\left(\rho_{\alpha}^{A}a_{\alpha}\right)-\frac{1}{V^{A}}\rho^{n}I^{n}V^{n}-\frac{\Delta t}{V^{A}}\sum_{s}h_{s}A_{s}T_{s}^{n}+ \\
\frac{\Delta t}{V^{A}}\sum_{s}\left[1-H\left(\rho_{h_2o}^{p}-\rho_{s,saturation}\right)\right]\cdot\max\left[\begin{array}{l}-\frac{\delta A_{s}\rho_{h_2ol}^{thermo}}{2\Delta t}I_{h_2o}\left(T_{s}\right), \\ h_{d}^{*}A_{s}\left(\rho_{h_2o}^{p}-\rho_{s,saturation}\right)I_{h_2o}\left(T_{s}\right)\end{array}\right]+ \\
\frac{\Delta t}{V^{A}}\sum_{s}H\left(\rho_{h_2o}^{p}-\rho_{s,saturation}\right)\cdot\max\left[\begin{array}{l}0, \\ a_{h_2o}h_{d}^{*}A_{s}\left(\rho_{h_2o}^{A}-\rho_{s,saturation}\right)\end{array}\right]
\end{array}
\right\}+
$$

$$
T^{A}\left\{
\begin{array}{l}
\sum_{\alpha}\left(\rho_{\alpha}^{A}b_{\alpha}\right)-\frac{\Delta t}{V^{A}}R_{h_2o}\left(S_{\rho,h_2o,condensation/vaporization}^{A}+S_{\rho,h_2o,mixture}^{A}\right)V^{n}+ \\
\frac{\Delta t}{V^{A}}\sum_{s}h_{s}A_{s}+\frac{\Delta t}{V^{A}}\sum_{s}H\left(\rho_{h_2o}^{p}-\rho_{s,saturation}\right)\cdot\max\left[\begin{array}{l}0, \\ b_{h_2o}h_{d}^{*}A_{s}\left(\rho_{h_2o}^{A}-\rho_{s,saturation}\right)\end{array}\right]
\end{array}
\right\}+
$$

$$
\left(T^{A}\right)^{2}\left\{
\begin{array}{l}
\sum_{\alpha}\left(\rho_{\alpha}^{A}c_{\alpha}\right)+ \\
\frac{\Delta t}{V^{A}}\sum_{s}H\left(\rho_{h_2o}^{p}-\rho_{s,saturation}\right)\cdot\max\left[\begin{array}{l}0, \\ c_{h_2o}h_{d}^{*}A_{s}\left(\rho_{h_2o}^{A}-\rho_{s,saturation}\right)\end{array}\right]
\end{array}
\right\}+
$$

$$
\left(T^{A}\right)^{3}\left\{
\begin{array}{l}
\sum_{\alpha}\left(\rho_{\alpha}^{A}d_{\alpha}\right)+ \\
\frac{\Delta t}{V^{A}}\sum_{s}H\left(\rho_{h_2o}^{p}-\rho_{s,saturation}\right)\cdot\max\left[\begin{array}{l}0, \\ d_{h_2o}h_{d}^{*}A_{s}\left(\rho_{h_2o}^{A}-\rho_{s,saturation}\right)\end{array}\right]
\end{array}
\right\}+
$$

$$
\left(T^{A}\right)^{4}\left\{
\begin{array}{l}
\sum_{\alpha}\left(\rho_{\alpha}^{A}e_{\alpha}\right)+ \\
\frac{\Delta t}{V^{A}}\sum_{s}H\left(\rho_{h_2o}^{p}-\rho_{s,saturation}\right)\cdot\max\left[\begin{array}{l}0, \\ e_{h_2o}h_{d}^{*}A_{s}\left(\rho_{h_2o}^{A}-\rho_{s,saturation}\right)\end{array}\right]
\end{array}
\right\}=0 \qquad (3\text{-}43)
$$

Once the mixture density and temperature are found, we can determine each of the energy source terms found in Equations (3-34) as follows:

$$R_{h_2o}T^A\left(S^A_{\rho,h_2o,condensation/vaporization}+S^A_{\rho,h_2o,mixture}\right)V^n=$$

$$-R_{h_2o}T^A\left\{\begin{array}{l}\sum_s^{evaporation}\left[1-H\left(\rho^p_{h_2o}-\rho_{s,saturation}\right)\right]\cdot\max\left[\begin{array}{c}-\frac{\delta^n_s A_s\rho^{thermo}_{h_2ol}}{2\Delta t},\\ h^*_d A_s\left(\rho^p_{h_2o}-\rho_{s,saturation}\right)\end{array}\right]+\\ \sum_s^{condensation}H\left(\rho^p_{h_2o}-\rho_{s,saturation}\right)\cdot\max\left[\begin{array}{c}0,\\ h^*_d A_s\left(\rho^A_{h_2o}-\rho_{s,saturation}\right)\end{array}\right]+\\ V^nC\left\{\begin{array}{l}\left[1-H\left(\rho^p_{h_2o}-\rho_{sat}(T^n,p_{sat})\right)\right]\cdot\max\left[\begin{array}{c}-\rho^n_{h_2ol}/10,\\ \rho^p_{h_2o}-\rho_{sat}(T^n,p_{sat})\end{array}\right]\\ +H\left(\rho^p_{h_2o}-\rho_{sat}(T^n,p_{sat})\right)\cdot\max\left[\begin{array}{c}0,\\ \rho^A_{h_2o}-\rho_{sat}(T^n,p_{sat})\end{array}\right]\end{array}\right\}\end{array}\right\} \qquad (3\text{-}44)$$

$$S^A_{I,convection}V^n=\sum_s h_s A_s\left(T^n_s-T^A\right) \qquad (3\text{-}45)$$

$$S^A_{I,condensation/vaporization}V^n=$$

$$-\sum_s^{evaporation}\left[1-H\left(\rho^p_{h_2o}-\rho_{s,saturation}\right)\right]\cdot\max\left[\begin{array}{c}-\frac{\delta^n_s A_s\rho^{thermo}_{h_2ol}}{2\Delta t},\\ h^*_d A_s\left(\rho^p_{h_2o}-\rho_{s,saturation}\right)\end{array}\right]$$

$$-\sum_s^{condensation}H\left(\rho^p_{h_2o}-\rho_{s,saturation}\right)\cdot\max\left[\begin{array}{c}0,\\ h^*_d A_s\left(\rho^A_{h_2o}-\rho_{s,saturation}\right)\end{array}\right]\quad . \qquad (3\text{-}46)$$

The total energy, $q_{s,total}$,, delivered to a given surface is

$$q_{s,total}=q_{s,convection}+q_{s,condensation/vaporization}=$$

$$h_s A_s \left(T^A - T_s^n\right) + \left[1 - H\left(\rho_{h_2o}^p - \rho_{s,saturation}\right)\right] \left\{ \begin{array}{l} \max\left[\begin{array}{l} -\dfrac{\delta_s^n A_s \rho_{h_2ol}^{thermo}}{2\Delta t}, \\ h_d^* A_s \left(\rho_{h_2o}^p - \rho_{s,saturation}\right) \end{array} \right] \cdot \\ \left[h_{fg} + I_{h_2o}\left(T_s^n\right)\right] \end{array} \right\}$$

$$+H\left(\rho_{h_2o}^p - \rho_{s,saturation}\right) \left\{ \begin{array}{l} \max\left[\begin{array}{l} 0, \\ h_d^* A_s \left(\rho_{h_2o}^A - \rho_{s,saturation}\right) \end{array} \right] \cdot \\ \left[h_{fg} + I_{h_2o}\left(T^A\right) - I_{h_2o}\left(T_s^n\right)\right] \end{array} \right\} \quad . \tag{3-47}$$

3.2.9 Liquid Droplet Depletion or "Rainout"

When the water liquid component is specified with user input, water droplets may be allowed to rain out if the accumulation of liquid water mass exceeds a specified loading value. The liquid water species mass Equation (3-6) with Equation (2-50) is

$$\frac{\rho_{h_2ol}^A V^A - \rho_{h_2ol}^n V^n}{\Delta t} = S_{\rho,h_2ol,rainout}^A V^n =$$

$$+V^n C_{h_2ol,mixture} \cdot \min\left[\begin{array}{l} 0, \\ \left(\rho_{h_2ol,\max} - \rho_{h_2ol}^A\right) \end{array} \right] \quad , \tag{3-48}$$

and the energy Equation (3-8) with Equation (2-51) is

$$\frac{\rho^A I^A V^A - \rho^n I^n V^n}{\Delta t} = S_{I,rainout}^A V^n =$$

$$V^n C_{h_2ol,mixture} \cdot \min\left[\begin{array}{l} 0, \\ \left(\rho_{h_2ol,\max} - \rho_{h_2ol}^A\right) \end{array} \right] \cdot I\left(T^A\right) \quad . \tag{3-49}$$

Equation (3-48) is only solved if the liquid density exceeds a specified value. If that should be true, then the advanced time-level liquid water density is

$$\rho_{h_2ol}^A = \frac{\rho_{h_2ol}^n V^n + \Delta t V^n C_{h_2ol,mixture} \rho_{h_2ol,\max}}{V^A + \Delta t V^n C_{h_2ol,mixture}} \quad , \tag{3-50}$$

which allows an easy evaluation of the rainout mass term

$$S_{\rho,h_2ol,rainout}^A V^n = V^n C_{h_2ol,mixture} \cdot \min\left[\begin{array}{l} 0, \\ \left(\rho_{h_2ol,\max} - \rho_{h_2ol}^A\right) \end{array} \right] \quad . \tag{3-51}$$

The time-advanced mixture temperature is computed from Equation (3-49) as

$$\begin{aligned}
&\left[\sum_\alpha \rho_\alpha^A a_\alpha - \rho^n I^n \frac{V^n}{V^A} - \Delta t \frac{V^n}{V^A} C_{h_2ol,mixture} \cdot \left(\rho_{h_2ol,\max} - \rho_{h_2ol}^A\right) a_{h_2ol}\right] \\
&+\left(T^A\right)\left[\sum_\alpha \rho_\alpha^A b_\alpha - \Delta t \frac{V^n}{V^A} C_{h_2ol,mixture} \cdot \left(\rho_{h_2ol,\max} - \rho_{h_2ol}^A\right) b_{h_2ol}\right] \\
&+\left(T^A\right)^2\left[\sum_\alpha \rho_\alpha^A c_\alpha - \Delta t \frac{V^n}{V^A} C_{h_2ol,mixture} \cdot \left(\rho_{h_2ol,\max} - \rho_{h_2ol}^A\right) c_{h_2ol}\right] \\
&+\left(T^A\right)^3\left[\sum_\alpha \rho_\alpha^A d_\alpha - \Delta t \frac{V^n}{V^A} C_{h_2ol,mixture} \cdot \left(\rho_{h_2ol,\max} - \rho_{h_2ol}^A\right) d_{h_2ol}\right] \\
&+\left(T^A\right)^4\left[\sum_\alpha \rho_\alpha^A e_\alpha - \Delta t \frac{V^n}{V^A} C_{h_2ol,mixture} \cdot \left(\rho_{h_2ol,\max} - \rho_{h_2ol}^A\right) d_{h_2ol}\right] = 0 \quad ,
\end{aligned} \tag{3-52}$$

which allows a straight forward evaluation of the rainout energy term

$$S_{I,rainout}^A V^n = V^n C_{h_2ol,mixture} \cdot \left(\rho_{h_2ol,\max} - \rho_{h_2ol}^A\right) \cdot I\left(T^A\right) \quad . \tag{3-53}$$

3.2.10 Hydrogen Combustion

3.2.10.1 One-Step Model

For the current one-step chemical kinetics model [Equation (2-85)], we see that three species–hydrogen, oxygen, and water vapor–are involved in this fundamental chemical balance. The mass Equation (3-6) for each of these species is

$$\begin{aligned}
\frac{\rho_{h_2}^A V^A - \rho_{h_2}^n V^n}{\Delta t} &= S_{\rho,h_2,combustion}^A V^n \\
\frac{\rho_{o_2}^A V^A - \rho_{o_2}^n V^n}{\Delta t} &= S_{\rho,o_2,combustion}^A V^n \\
\frac{\rho_{o_2}^A V^A - \rho_{o_2}^n V^n}{\Delta t} &= S_{\rho,o_2,combustion}^A V^n
\end{aligned} \tag{3-54}$$

and the energy Equation (3-8) is

$$\frac{\rho^A I^A V^A - \rho^n I^n V^n}{\Delta t} = S_{I,combustion}^A V^n \quad . \tag{3-55}$$

We use Equations (2-86), (2-89), (2-91), and (2-95) to obtain the right-hand sides of the above equations, with

$$\frac{d\rho_\alpha}{dt} = M_\alpha \frac{dc_\alpha}{dt} \quad . \tag{3-56}$$

If the mixture is fuel lean, $c_{h_2} < 2 \cdot c_{o_2}$, we solve Equation (2-93) and arrive at

$$-\frac{1}{2}\frac{c_{h_2}^A - c_{h_2}^n}{\Delta t} = \dot{\omega} = \frac{k(T)c_{o_2}^n c_{h_2}^n}{1 + 2\Delta t k(T) c_{o_2}^n} \quad , \tag{3-57}$$

and if the mixture is fuel rich, $c_{h_2} > 2 \cdot c_{o_2}$, we solve Equation (2-94) and arrive at

$$-\frac{c_{o_2}^A - c_{o_2}^n}{\Delta t} = \dot{\omega} = \frac{k(T)c_{o_2}^n c_{h_2}^n}{1 + \Delta t k(T) c_{h_2}^n} \quad , \tag{3-58}$$

which gives the mass source terms for Equation (3-54)

$$\begin{aligned} S_{\rho,h_2,combustion}^A &= -2M_{h_2}\dot{\omega} \\ S_{\rho,o_2,combustion}^A &= -M_{o_2}\dot{\omega} \\ S_{\rho,h_2o,combustion}^A &= -2M_{h_2o}\dot{\omega} \end{aligned} \tag{3-59}$$

and the energy source term

$$S_{I,combustion}^A = C_c \cdot \dot{\omega} \quad . \tag{3-60}$$

Note that the solutions for the reaction rate, $\dot{\omega}$, in both Equations (3-57) and (3-58) contain the rate constant described by Equation (2-91), which is only a function of the mixture temperature. When computing the chemical kinetics of this model, the temperature is held constant, but it is this rate constant that provides the key to modeling ignitors. When it is determined from the user-supplied input (see NUREG/CR-6570, Vol. 2) to activate an ignitor, the rate constant as described in Section 2.10 becomes which effectively ignites the flammable mixture.

3.2.10.2 Two-Step Model

For this model, as for the one-step model, the three species hydrogen, oxygen, and water vapor are involved in the chemical balance. The mass Equation (3-6) for each of these species is

$$\frac{\rho_{h_2}^A V^A - \rho_{h_2}^n V^n}{\Delta t} = S_{\rho,h_2,combustion}^A V^n$$
$$\frac{\rho_{o_2}^A V^A - \rho_{o_2}^n V^n}{\Delta t} = S_{\rho,o_2,combustion}^A V^n \qquad (3\text{-}61)$$
$$\frac{\rho_{h_2o}^A V^A - \rho_{h_2o}^n V^n}{\Delta t} = S_{\rho,h_2o,combustion}^A V^n$$

and the energy Equation (3-8) is

$$\frac{\rho^A I^A V^A - \rho^n I^n V^n}{\Delta t} = S_{I,combustion}^A V^n \ . \qquad (3\text{-}62)$$

When the induction parameter is greater than 1, meaning that energy is released, the mass source term for Equation (3-54) for hydrogen is calculated in two different ways depending on the regime the combustion is in.

If the local turbulent Reynolds number is smaller than a critical value, the combustion takes place in a quasilaminar regime. Hydrogen is burned in a stepwise linear way. The rate depends on the energy release time and the subcycle time step:

$$-\frac{c_{h_2}^A - c_{h_2}^n}{\Delta t_{sub}} = \dot{\omega} = \frac{c_{h_2,beforecombustion}}{t_{rel}} \ . \qquad (3\text{-}63)$$

If the local turbulent Reynolds number is greater than a critical value, the combustion takes place in a fully turbulent regime. Assuming the Damkoehler number (ratio of induction time to turbulent time) is greater than the ignition/extinction criterion, the mass source term for hydrogen is calculated as follows:

$$-\left(c_{h_2}^A - c_{h_2}^n\right) = \frac{2 \cdot A}{\tau_t} \cdot c_{\lim}^n \ . \qquad (3\text{-}64)$$

The resulting mass source terms for oxygen and water vapor are

$$S_{\rho,o_2,combustion}^A = -\frac{1}{2} M_{o_2} \dot{\omega}$$
$$S_{\rho,h_2o,combustion}^A = -M_{h_2o} \dot{\omega} \qquad (3\text{-}65)$$

and the energy source term

$$S_{I,combustion}^A = C_c \cdot \dot{\omega} \ . \qquad (3\text{-}66)$$

In case the Damkoehler number is smaller than D_{ie}, no hydrogen is burned, so the mass and energy source terms are zero.

If the energy is released in a quasilaminar regime in a stepwise linear way, the combustion timestep is determined by dividing the minimum value of the energy release time by the number of timesteps used for energy release:

$$\Delta t_{combustion} = \frac{t_{rel,\min}}{relstep} \quad . \tag{3-67}$$

The value of *relstep* can be input by the user; the default is 1.

If the combustion time step is very small and would determine the timestep of the next cycle, the combustion will be subcycled with this timestep to allow for a larger timestep in the fluid dynamics routines.

3.2.11 Hydrogen Recombination

For the hydrogen recombination chemical kinetics, three species–hydrogen, oxygen, and water vapor–are involved in the fundamental chemical balance. The mass Equation (3-6) for each of these species is

$$\begin{aligned} \frac{\rho_{h_2}^A V^A - \rho_{h_2}^n V^n}{\Delta t} &= S_{\rho,h_2,recombination}^A V^n \\ \frac{\rho_{o_2}^A V^A - \rho_{o_2}^n V^n}{\Delta t} &= S_{\rho,o_2,recombination}^A V^n \\ \frac{\rho_{h_2o}^A V^A - \rho_{h_2o}^n V^n}{\Delta t} &= S_{\rho,h_2o,recombination}^A V^n \end{aligned} \tag{3-68}$$

and the energy Equation (3-8) is

$$\frac{\rho^A I^A V^A - \rho^n I^n V^n}{\Delta t} = S_{I,recombination}^A V^n \quad . \tag{3-69}$$

For each of the recombiner models currently modeled in GASFLOW-MPI (see Section 2.9), we develop a reaction rate, $\dot{\omega}$, for the consumption of hydrogen. We review those reaction rates here.

For the NIS recombiner box (see Section 2.9.1), the hydrogen recombination rate is given in Equation (2-123).

For the Siemens recombiner box (see Section 2.9.2), the hydrogen recombination rate is given in Equation (2-128).

For the Siemens correlation for Type FR-90/1 Recombiner (see Section 2.9.3), the hydrogen recombination rate is given in Equation (2-135).

For the GRS correlation (see Section 2.9.4), the hydrogen recombination rate is given in Equation (2-138).

In each of these cases, the change of hydrogen concentration is computed from Equation (3-57) in the form

$$-\frac{1}{2}\frac{c_{h_2}^A - c_{h_2}^n}{\Delta t} = \dot{\omega} \tag{3-70}$$

or the recombination source terms for the species mass equations

$$\begin{aligned} S_{\rho,h_2,recombination}^A &= -2M_{h_2}\dot{\omega} \\ S_{\rho,o_2,recombination}^A &= -M_{o_2}\dot{\omega} \\ S_{\rho,h_2o,recombination}^A &= -2M_{h_2o}\dot{\omega} \end{aligned} \tag{3-71}$$

and the energy source term

$$S_{I,recombination}^A = C_c \cdot \dot{\omega} \quad . \tag{3-72}$$

3.3 Phase B: Implicit Pressure Iteration Phase

3.3.1 Three-Dimensional Developments

In this phase, an implicit evaluation of the time-advanced densities, velocities, pressure, and specific internal energy fields is achieved. The purpose of this phase is to compute time-advanced pressures to allow calculations of low-speed (low-Mach-number) flows without any time-step restrictions from the fluid sound speeds. The following argument (Ref. 3-1) explains the need for this step.

In an explicit method, pressure forces can be transmitted only one cell each time step, that is, cells exert pressure forces only on neighboring cells. When the time step is chosen so large that sound waves should travel more than one cell, the one cell limitation is clearly inaccurate and a catastrophic instability develops. The instability arises because the explicit pressure gradients lead to excessive cell compressions or expansions when multiplied by too large a time step. This then leads to larger pressure gradients the next cycle, which try to reverse the previous excesses, but since the time step is too large the reversal is also too large and the process repeats itself with a rapidly increasing amplitude. The over response to pressure gradients in this fashion is eliminated by using time-advanced pressure gradients, for then cells cannot compress or expand to the point where gradients are reversed.

In this phase, the mixture differential equations for cell volume [Equation (2–8)], mass [Equation (2-7)], momentum [Equation (2-13)], and energy [Equation (2-14)] are

$$\frac{V^B - V^A}{\Delta t} = V^n \nabla \cdot \left[(A\mathbf{u})^B - (A\mathbf{u})^n \right] \quad , \tag{3-73}$$

$$\frac{\rho^B V^B - \rho^A V^A}{\Delta t} = 0 \quad , \tag{3-74}$$

$$\frac{\rho_m^B V_m^B \mathbf{u}^B - \rho_m^A V_m^A \mathbf{u}^A}{\Delta t} = V_m^n \left[-\nabla(p^B - p^n) - (A\mathbf{D}^B - A\mathbf{D}^A) \right] \quad , \tag{3-75}$$

and

$$\frac{\rho^B V^B I^B - \rho^A V^A I^A}{\Delta t} = -V^n p^n \nabla \cdot \left[(A\mathbf{u})^B - (A\mathbf{u})^n \right] \quad . \tag{3-76}$$

The equation of state for the mixture [Equation (2-25)] may be written as

$$T^B = \frac{\theta^B p^B V^B}{\rho^B V^B \sum_{\alpha}^{\alpha \neq h_2 ol} x_\alpha^B R_\alpha} \quad , \tag{3-77}$$

and with the approximation for the mixture specific heat at constant volume

$$c_v(T^A) \cong \frac{\left(I^B - I^A\right)}{\left(T^B - T^A\right)} \tag{3-78}$$

we present an equation set, Equations (3-73) through (3-78), that is coupled with seven linear and one nonlinear [Equation (3-77)] algebraic equation in eight unknowns (V^B, ρ^B, u_x^B, u_y^B, u_z^B, p^B, I^B, and T^B). We are able to reduce this equation set to a single Poisson equation involving the pressure change as the dependent variable.

The Poisson equation is solved for the change in the pressure field, and from the pressure change, the volume, velocity, density, and internal energy fields are found by back substitution.

The Poisson equation is derived as follows:

(1) The left-hand side of the conservation of energy Equation (3-76) is manipulated using the conservation of mass Equation (3-74), the equation of state (3-77), and the specific heat approximation (3-78) to yield

$$\frac{\theta^A c_v(T^A)\left(p^B V^B - p^A V^A\right)}{\Delta t \sum_{\alpha}^{\alpha \neq h_2 ol} x_\alpha^A R_\alpha} = -V^n p^n \nabla \cdot \left[(A\mathbf{u})^B - (A\mathbf{u})^n \right] \quad . \tag{3-79}$$

(2) The right-hand side of Equation (3-79) can be modified using the conservation of volume Equation (3-73); and, in addition, $p^A V^B$ is subtracted and added to the numerator of the left-hand side to give

$$\frac{\theta^A c_v(T^A)\left(p^B V^B - p^A V^B + p^A V^B - p^A V^A\right)}{\sum_{\alpha}^{\alpha \neq h_2 ol} x_\alpha^A R_\alpha} = -p^n \left(V^B - V^A\right) \quad . \tag{3-80}$$

(3) Rearranging Equation (3-84) yields the following:

$$p^B - p^A = -\left[\frac{p^n \sum\limits_{\alpha}^{\alpha \neq h_2 ol} x_\alpha^A R_\alpha}{\theta^A c_v(T^A)} + p^A \right] \frac{\left(V^B - V^A\right)}{V^B} \quad . \tag{3-81}$$

(4) This equation can be linearized by rearranging $\left(V_s^B - V_s^A\right)/V_s^B$ and then applying a binomial series to obtain

$$\frac{\left(V^B - V^A\right)}{V^B} \cong \frac{\left(V^B - V^A\right)}{V^A} \quad , \tag{3-82}$$

provided $\frac{\left(V^B - V^A\right)}{V^A} << 1$, to produce

$$p^B - p^A = -\left[\frac{p^n \sum\limits_{\alpha}^{\alpha \neq h_2 ol} x_\alpha^A R_\alpha}{\theta^A c_v(T^A)} + p^A \right] \frac{\left(V^B - V^A\right)}{V^A} \quad . \tag{3-83}$$

(5) Adding and subtracting $(A\mathbf{u})^A$ inside the divergence operator in Equation (3-73) yields

$$V^B - V^A = \Delta t V^n \nabla \cdot \left[(A\mathbf{u})^B - (A\mathbf{u})^A + (A\mathbf{u})^A - (A\mathbf{u})^n \right] \quad . \tag{3-84}$$

(6) Equation (3-84) can be substituted into Equation (3-83) and, after some algebra, yields

$$\frac{V^A}{\Delta t}\left(p^B - p^A\right) = -V^n \{\ \} \left\{ \nabla \cdot \left[(A\mathbf{u})^B - (A\mathbf{u})^A \right] + \nabla \cdot \left[(A\mathbf{u})^A - (A\mathbf{u})^n \right] \right\} \quad , \tag{3-85}$$

where

$$\{\ \} = \left[\frac{p^n \sum\limits_{\alpha}^{\alpha \neq h_2 ol} x_\alpha^A R_\alpha}{\theta^A c_v(T^A)} + p^A \right] \quad .$$

(7) Making use of Equation (3-74), Equation (3-75) can be cast into the following form:

$$(A\mathbf{u})^B - (A\mathbf{u})^A = -\frac{\mathbf{A} \Delta t V_m^n \nabla (p^B - p^n)}{\langle\ \rangle} \quad , \tag{3-86}$$

where

$$\langle\ \rangle = \rho_m^A V_m^A \left(1+\frac{\Delta t \mathbf{C}\left|\mathbf{u}^n\right|}{2\Delta \mathbf{x}}\right) \quad .$$

(8) Introducing $\delta p = p^B - p^n$ into Equations (3-85) and (3-86) and eliminating $(A\mathbf{u})^B - (A\mathbf{u})^A$ between the two equations yields

$$\Delta t^2 \nabla \cdot \left[\frac{\mathbf{A} V_m^n \nabla \delta p}{\langle\ \rangle}\right] - \frac{V^A}{V^n\{\ \}}\delta p =$$

$$\frac{V^A}{V^n\{\ \}}(p^n - p^A) + \Delta t \nabla \cdot \left[(A\mathbf{u})^B - (A\mathbf{u})^n\right] \quad , \tag{3-87}$$

which is second order and linear in δp. To solve this Poisson pressure change equation, the Portable Extensible Toolkit for Scientific Computation (PETSc) developed by Argonne National Laboratory is used. In GASFLOW-MPI 1.0, we use the Krylov method conjugate gradients plus block Jacobi as the pre-conditioner.

This implicit solution of the pressure equation allows for greater efficiency than a purely explicit calculation with reduced time steps. The numerical stability achieved permits pressure waves to traverse more than one computational cell in a time step.

In practice, after solving Equation (3-87) for the pressure change, δp, we evaluate $\mathbf{u}^B$ from Equation (3-86), since $\delta p = p^B - p^n$, V^B from Equation (3-73), the time-advanced density ρ^B from Equation (3-74), ρ_α^B from

$$\frac{\rho_\alpha^B V^B - \rho_\alpha^A V^A}{\Delta t} = 0 \quad , \tag{3-88}$$

p^B from Equation (3-81), T^B from Equation (3-77), and finally I^B from a direct evaluation of Equations (2-19) as

$$I^B = \sum_\alpha \left(\frac{\rho_\alpha^B}{\rho^B} a_\alpha\right) +$$

$$T^B \left\langle \sum_\alpha \left(\frac{\rho_\alpha^B}{\rho^B} b_\alpha\right) + T^B \left\{ \sum_\alpha \left(\frac{\rho_\alpha^B}{\rho^B} c_\alpha\right) + T^B \left[\sum_\alpha \left(\frac{\rho_\alpha^B}{\rho^B} d_\alpha\right) + T^B \sum_\alpha \left(\frac{\rho_\alpha^B}{\rho^B} e_\alpha\right) \right] \right\} \right\rangle \quad . \tag{3-89}$$

3.3.2 Ventilation System Developments

The one-dimensional duct Phase B solution is the same as that used in the three-dimensional blocks, except for the inclusion of the momentum source terms for the blower operation and the change in momentum flux. These two momentum source terms are included into Equation (3-90) as explicit source terms in the difference between the Phase A and Phase B,

$$\frac{\rho_m^B V_m^B \mathbf{u}^B - \rho_m^A V_m^A \mathbf{u}^A}{\Delta t} = V_m^n \left[-\nabla(p^B - p^n) - (A\mathbf{D}^B - A\mathbf{D}^A) + \rho_m^n g_c H_b A - \sum_f (\rho A \mathbf{u}\mathbf{u})_f^n \Delta S_f \right] . \tag{3-90}$$

In addition, the total inclusion of the flow loss in the Phase B solution is consistent with the internal structure drag in the three-dimensional blocks. With these two modifications, the development of the Phase B pressure equation follows the three-dimensional Phase B development of the pressure equation.

3.4 Phase C: Rezone Phase

3.4.1 Multidimensional Developments

The third phase explicitly performs all the advective flux calculations, repartitioning the dependent variables onto the original mesh. The superscript *n+1* (and not *C*) is used to indicate that this Eulerian rezone phase completes the spatiotemporal integration of the equations of motion from time-level *n* to time-level *n+1*. Recognizing that $V_s^{n+1} = V_s^n$ because of the exact remap of the Lagrangian mesh onto the original Eulerian mesh, the finite volume equations for mass, momentum, and energy advection, respectively, are as follows:

$$\frac{\rho_\alpha^{n+1} V^n - \rho_\alpha^B V^B}{\Delta t} = -\sum_f (\rho_\alpha A \mathbf{u})_f^B \Delta S_f \quad , \tag{3-91}$$

$$\frac{\rho_m^{n+1} V_m^n \mathbf{u}^{n+1} - \rho_m^B V_m^B \mathbf{u}^B}{\Delta t} = -\sum_f (\rho A \mathbf{u}\mathbf{u})_f^B \Delta S_f - \sum_f (\mathbf{D}_d \mathbf{A})_f^{n+1} \Delta S_f \quad , \tag{3-92}$$

and

$$\frac{\rho^{n+1} V^n I^{n+1} - \rho^B V^B I^B}{\Delta t} = -\sum_f (\rho I A \mathbf{u})_f^B \Delta S_f \quad . \tag{3-93}$$

Decoupling the rezoning step from the rest of the physics computations facilitates the implement-tation of different numerical advection algorithms. We have implemented both a first-order donor cell algorithm and a second order van Leer with limiting (Ref. 3-8) to evaluate the right-hand sides of Equations (3-91) through (3-93).

To illustrate the van Leer algorithm, we first expand the right-hand side of the species mass, momenta, or energy in Equations (3-91) through (3-93) over the finite-volume control surfaces as

$$\sum_f (\phi Au)_f^B \Delta S_f = \begin{cases} \langle \phi Au \rangle_E^B \delta y \delta z - \langle \phi Au \rangle_W^B \delta y \delta z + \\ \langle \phi Av \rangle_N^B \delta x \delta z - \langle \phi Av \rangle_S^B \delta x \delta z + \\ \langle \phi Aw \rangle_T^B \delta x \delta y - \langle \phi Aw \rangle_B^B \delta x \delta y \end{cases} , \tag{3-94}$$

where ϕ is, respectively, ρ_α , $\rho\mathbf{u}$, and ρI . We remind the reader that *A* is the fractional area open for flow on that particular surface.

The East (E) surface of computation volume (I, J, K) corresponding with the mass and energy equations coincides with the indexing notation of *i*+1/2, where the special notation $\langle\ \rangle$ is defined:

$$\langle \phi Au \rangle_{i+\frac{1}{2}}^B = \begin{cases} A_{i+\frac{1}{2}} u_{i+\frac{1}{2}}^B \left\{ \phi_i^B + \frac{1}{2}\left[\delta x_i - u_{i+\frac{1}{2}}^B \delta t \right] \overline{\left(\frac{\partial \phi^B}{\partial x} \right)_i} \right\} ; u_{i+\frac{1}{2}}^B \geq 0 \\ A_{i+\frac{1}{2}} u_{i+\frac{1}{2}}^B \left\{ \phi_{i+1}^B + \frac{1}{2}\left[\delta x_{i+1} + u_{i+\frac{1}{2}}^B \delta t \right] \overline{\left(\frac{\partial \phi^B}{\partial x} \right)_{i+1}} \right\} ; u_{i+\frac{1}{2}}^B < 0 \end{cases} . \tag{3-95}$$

We make use of central differencing to evaluate the average slope in Equation (3-95) when the fluid velocity at face *i*+1/2 is positive as

$$\overline{\left(\frac{\partial \phi^B}{\partial x} \right)_i} = \overline{S_i} \cong \frac{\phi_{i+1}^B - \phi_{i-1}^B}{\delta x_{i+\frac{1}{2}} + \delta x_{i-\frac{1}{2}}} , \tag{3-96}$$

which is interpreted schematically in Figure 3-6.

When the fluid velocity at face *i*+1/2 is negative, the average slope in Equation (3-95) is given by

$$\overline{\left(\frac{\partial \phi^B}{\partial x} \right)_{i+1}} = \overline{S_{i+1}} \cong \frac{\phi_{i+2}^B - \phi_i^B}{\delta x_{i+\frac{3}{2}} + \delta x_{i+\frac{1}{2}}} , \tag{3-97}$$

which is interpreted schematically in Figure 3-7.

Note that when the average slope in Equation (3-95) is identically zero, the algorithm reduces to the classical first-order donor cell or Godunov's first-order scheme:

$$(\phi Au)_{i+\frac{1}{2}}^B = \begin{cases} A_{i+\frac{1}{2}} u_{i+\frac{1}{2}}^B \phi_i^B ; u_{i+\frac{1}{2}}^B \geq 0 \\ A_{i+\frac{1}{2}} u_{i+\frac{1}{2}}^B \phi_{i+1}^B ; u_{i+\frac{1}{2}}^B < 0 \end{cases} . \tag{3-98}$$

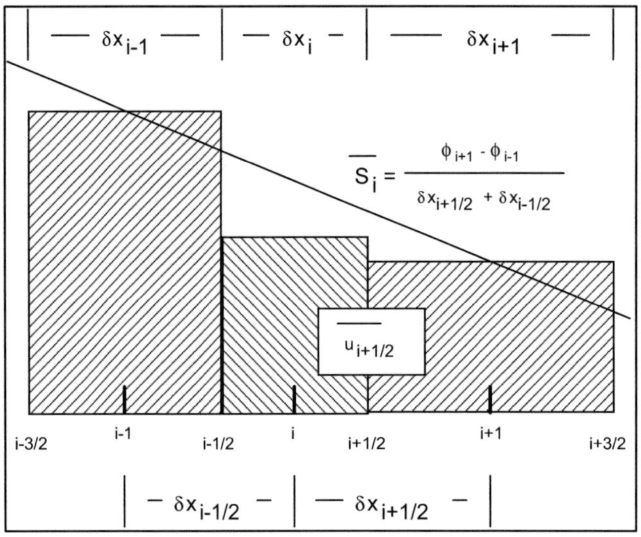

Figure 3-6 Schematic diagram of van Leer second-order advection algorithm when face velocity is positive

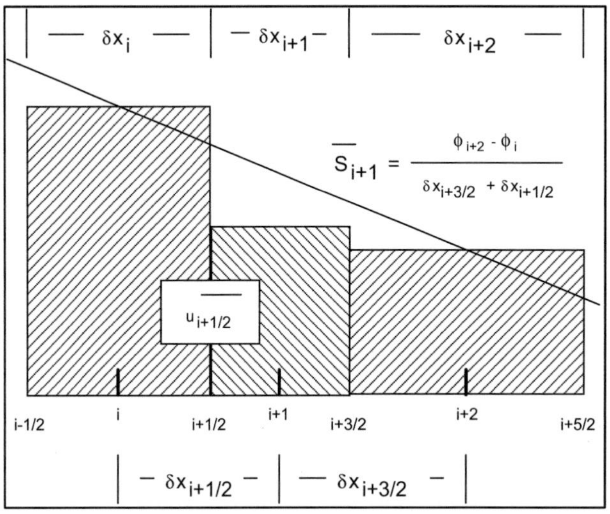

Figure 3-7 Schematic diagram of van Leer second-order advection algorithm when face velocity is negative

By limiting the value of the slope in various situations (which we will discuss below) we can impose the monotonicity condition. Basically, the monotonicity condition states that when the initial conditions for a particular variable are monotone, the time-advanced values are also monotone. In other words, if ϕ_i^B lies between ϕ_{i+1}^B and ϕ_{i-1}^B, then ϕ_i^{n+1} must lie between ϕ_{i+1}^{n+1} and ϕ_{i-1}^{n+1}.

The idea of limiting the slope is shown in Figure 3-8, Figure 3-9, and Figure 3-10, which show the slope distribution of ϕ. In Figure 3-8, the slope of the variable between *i*-1/2 and *i*+1/2 lies outside the zero slope values for *i*-1 and *i*+1. We limit the value of the slope in this case so that the actual slope used when computing Equation (3-95) is the heavy dark slope shown for *i* in Figure 3-8. In Figure 3-9, the mesh value at *i* reaches an extremum; in this case, the slope is reduced to zero. Also, if the slope at *i* doesn't agree with the trend of the adjacent slopes as shown in Figure 3-10, the slope is reduced to zero.

Based upon this limiter model, we retain or reduce the slope as computed by Equation (3-96) for positive fluid velocity at *i*+1/2 by

$$\left(\bar{S}_i\right)_{monotone} = \left\{ \begin{array}{c} sign\left(\phi_{i+1}^B - \phi_{i-1}^B\right) \cdot \min\left(\left(\left|\dfrac{\phi_i^B - \phi_{i-1}^B}{\delta x_{i-\frac{1}{2}}}\right|\right), \left(\left|\dfrac{\phi_{i+1}^B - \phi_{i-1}^B}{\delta x_{i+\frac{1}{2}} + \delta x_{i-\frac{1}{2}}}\right|\right), \left(\left|\dfrac{\phi_{i+1}^B - \phi_i^B}{\delta x_{i+\frac{1}{2}}}\right|\right) \right); \\ if: sign\left(\phi_i^B - \phi_{i-1}^B\right) = sign\left(\phi_{i+1}^B - \phi_{i-1}^B\right) = sign\left(\phi_{i+1}^B - \phi_i^B\right) \\ 0;\ otherwise \end{array} \right\} \quad (3\text{-}99)$$

and Equation (3-97) for negative fluid velocity at *i*+1/2 by

$$\left(\bar{S}_{i+1}\right)_{monotone} = \left\{ \begin{array}{c} sign\left(\phi_{i+2}^B - \phi_i^B\right) \cdot \min\left(\left(\left|\dfrac{\phi_{i+1}^B - \phi_i^B}{\delta x_{i+\frac{1}{2}}}\right|\right), \left(\left|\dfrac{\phi_{i+2}^B - \phi_i^B}{\delta x_{i+\frac{3}{2}} + \delta x_{i+\frac{1}{2}}}\right|\right), \left(\left|\dfrac{\phi_{i+2}^B - \phi_{i+1}^B}{\delta x_{i+\frac{3}{2}}}\right|\right) \right); \\ if: sign\left(\phi_{i+1}^B - \phi_i^B\right) = sign\left(\phi_{i+2}^B - \phi_i^B\right) = sign\left(\phi_{i+2}^B - \phi_{i+1}^B\right) \\ 0;\ otherwise \end{array} \right\} . \quad (3\text{-}100)$$

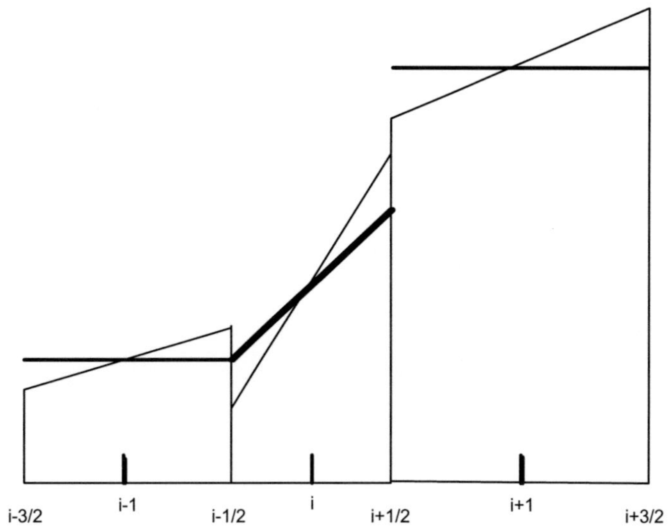

Figure 3-8 A monotonicity condition shown by plotting the slope distribution of ϕ (slope of i lies outside the zero slope values for i - 1 and i + 1).

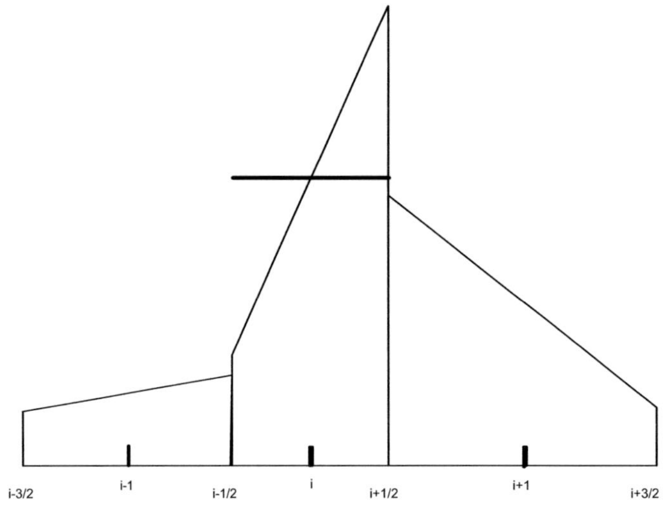

Figure 3-9 A monotonicity condition shown by plotting the slope distribution of ϕ (slope of i is reduced to zero when its mesh value reaches an extremum)

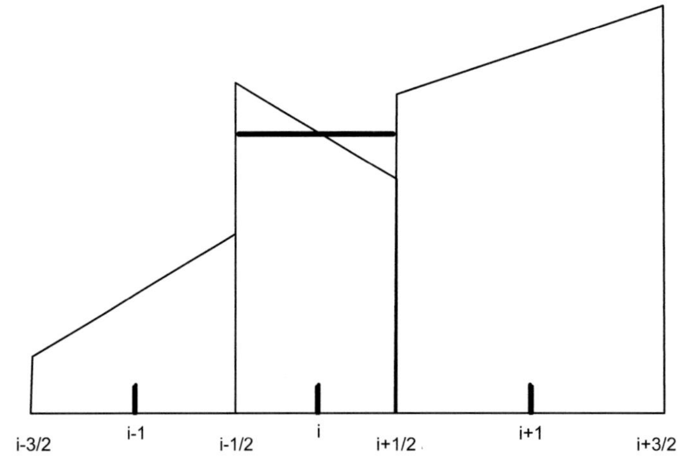

Figure 3-10 A monotonicity condition shown by plotting the slope distribution ϕ (slope of i is reduced to zero when it doesn't agree with the trend of the adjacent slopes)

The actual implemented advection algorithm is therefore

$$\langle \phi A u \rangle^B_{i+\frac{1}{2}} = \begin{cases} A_{i+\frac{1}{2}} u^B_{i+\frac{1}{2}} \left\{ \phi^B_i + \frac{1}{2}\left[\delta x_i - u^B_{i+\frac{1}{2}} \delta t \right] \left(\overline{S_i} \right)_{monotone} \right\}; u^B_{i+\frac{1}{2}} \geq 0 \\ A_{i+\frac{1}{2}} u^B_{i+\frac{1}{2}} \left\{ \phi^B_{i+1} + \frac{1}{2}\left[\delta x_{i+1} + u^B_{i+\frac{1}{2}} \delta t \right] \left(\overline{S_{i+1}} \right)_{monotone} \right\}; u^B_{i+\frac{1}{2}} < 0 \end{cases} . \quad (3\text{-}101)$$

Equation (3-91) can be solved using the structural drag function Equation (3-3) to yield

$$\left(\rho_m \mathbf{u}\right)^{n+1} = \frac{\rho_m^B V_m^B \mathbf{u}^B - \Delta t \sum_f \left(\rho A \mathbf{uu}\right)_f^B \Delta S_f}{V_m^n + \Delta t \mathbf{A} \frac{1}{2} \mathbf{C}_D \left|\mathbf{u}^B\right| S_f} \quad . \tag{3-102}$$

Phase C is completed by computing the total density

$$\rho^{n+1} = \sum_\alpha \rho_\alpha^{n+1} \tag{3-103}$$

T^{n+1} from inverting the polynomial of up to 4th degree

$$\left(\sum_\alpha \rho_\alpha^{n+1} a_\alpha - \rho^{n+1} I^{n+1}\right) + (T^{n+1}) \sum_\alpha \rho_\alpha^{n+1} b_\alpha +$$

$$(T^{n+1})^2 \sum_\alpha \rho_\alpha^{n+1} c_\alpha + (T^{n+1})^3 \sum_\alpha \rho_\alpha^{n+1} d_\alpha + (T^{n+1})^4 \sum_\alpha \rho_\alpha^{n+1} e_\alpha = 0 \tag{3-104}$$

and p^{n+1} from

$$p^{n+1} = \frac{T^{n+1} \sum_\alpha^{\alpha \neq h_2 ol} R_\alpha \rho_\alpha^{n+1}}{\theta^{n+1}} \tag{3-105}$$

with

$$\theta^{n+1} = 1 - \frac{\rho_{h_2 ol}^{n+1}}{\rho_{h_2 ol}^{thermo}} \quad . \tag{3-106}$$

3.4.2 Ventilation System Developments

Again, the one-dimensional duct Phase C solution is the same as the three-dimensional Phase C solution with the following exception. Since the change in momentum flux for the one-dimensional ducts is included in the Phase A and Phase B solutions for one-dimensional ducts, there is no Phase C momentum equation solution for one-dimensional ducts. The velocity obtained from the Phase B solution for one-dimensional ducts is the new time velocity, which is not modified in Phase C. The mass and energy equations for Phase C for one-dimensional ducts are the same as those for the three-dimensional blocks.

3.5 Turbulence Transport

The final fluid dynamics task in each computational cycle is the explicit updating of turbulence variables.

3.5.1 Algebraic Model

The new time-level turbulent viscosity follows directly from Equation (2-65):

$$\mu^{n+1} = C_\mu \rho^{n+1} \left\{ 0.1 \cdot \frac{1}{2} \left[\left(u^{n+1}\right)^2 + \left(v^{n+1}\right)^2 + \left(w^{n+1}\right)^2 \right] \right\}^{1/2} \quad . \tag{3-107}$$

3.5.2 κ-ε Model

The new time-level products $\rho\kappa$ and $\rho\varepsilon$ are computed from the discrete versions of Equations (2-68) and (2-69):

$$\frac{(\rho\kappa)^{n+1} V^n - (\rho\kappa)^n V^n}{\Delta t} = \sum_f \begin{bmatrix} -\rho\kappa\mathbf{u} + \\ \left(\frac{\mu}{\sigma_\kappa}\nabla\kappa\right) + \\ \tau : \nabla\mathbf{u} \end{bmatrix}_f^n A_f \Delta S_f + V^n \left[\mu\alpha\mathbf{g}\cdot\nabla T - \rho\varepsilon + \kappa S_\kappa\right]^n \tag{3-108}$$

and

$$\frac{(\rho\varepsilon)^{n+1} V^n - (\rho\varepsilon)^n V^n}{\Delta t} = \sum_f \begin{bmatrix} -\rho\varepsilon\mathbf{u} + \\ \left(\frac{\mu}{\sigma_\varepsilon}\nabla\varepsilon\right) + \\ C_1 \frac{\varepsilon}{\kappa} \tau : \nabla\mathbf{u} \end{bmatrix}_f^n A_f \Delta S_f +$$

$$V^n \left[C_1 \frac{\varepsilon}{\kappa} \mu\alpha\mathbf{g}\cdot\nabla T - C_1 \rho \frac{\varepsilon^2}{\kappa} + \varepsilon S_\varepsilon \right]^n \quad . \tag{3-109}$$

The turbulent viscosity then is computed from Equation (2-70):

$$\mu^{n+1} = \frac{C_\mu \rho^{n+1} \left(\kappa^{n+1}\right)^2}{\varepsilon^{n+1}} \quad . \tag{3-110}$$

3.6 Time-Step Controls

The new time step is controlled by checking the entire computational mesh for a material velocity Courant condition, the diffusion stability limit, the ignition of hydrogen, the pressure iteration count, and the maximum time step allowed by input.

3.6.1 Courant Condition for Material Velocity

The Courant material velocity time-step limitation is computed as the minimum limit on the entire computational mesh by

$$\delta t_{convection} = \frac{1}{4 \cdot \max\left\{ \dfrac{u_i}{\delta x_i}, \dfrac{v_j}{\delta y_j}, \dfrac{w_k}{\delta z_k}, \varepsilon_{convection} \right\}} , \tag{3-111}$$

where $\varepsilon_{convection}$ = 10^{-10}. Note that for a zero velocity field, the convective time-step limit is roughly the reciprocal of $\varepsilon_{convection}$, which is a large value. The coefficient 4 effectively limits material convection to one-fourth of the smallest computational cell in the mesh. We have experimented with this coefficient by relaxing its value to nearly 1. The solution remained very stable; however, for accuracy considerations we recommend the default value of 4.

3.6.2 Diffusional Condition for Mass, Momentum, and Energy

The diffusional time-step limitation is calculated every time step for the minimum values on the computational mesh by

$$\delta t_{diffusion} = \frac{1}{4 \cdot \lambda_{i,j,k} \cdot \left(\dfrac{1}{\delta x_i^2} + \dfrac{1}{\delta y_j^2} + \dfrac{1}{\delta z_k^2} \right)} , \tag{3-112}$$

where

$$\lambda_{i,j,k} = \max\left[\left(D_{apparent} \right)_{i,j,k}, \left(\frac{\mu_{apparent}}{\rho} \right)_{i,j,k}, \left(\frac{\phi_{apparent}}{\rho \cdot c_p} \right)_{i,j,k}, \varepsilon_{diffusion} \right] \tag{3-113}$$

and $\varepsilon_{diffusion}$ = 10^{-10}. Formally, the coefficient 4 in Equation (3-111) can be relaxed to 2, but we have chosen to be conservative in our calculation.

3.6.3 Ignition Conditions

When an ignition occurs, the energy of combustion dominates the internal energy Equation (3-8). We have determined an empirical relationship to limit the time step at ignition periods. Our calculation of the ignition time step is based upon the change of energy during ignition:

$$\delta t_{combustion} = \max\left[5 \cdot 10^{-4}, \frac{5 \cdot 10^{-5} \cdot I^n}{\left(I^A - I^n \right)} \right] . \tag{3-114}$$

3.6.4 Pressure Iteration Considerations

The user can define a reference pressure iteration variable (default value: itmax = 20) (Ref. 3-9) . This iteration time-step control is formulated as

$$\delta t_{iteration} = \left[1.02 - H(iteration - reference) \cdot 0.04\right] \cdot \delta t \quad , \tag{3-115}$$

where $H(iteration - reference)$ is the usual Heaviside function. This time-step control allows the time step to target a reference iteration value by either increasing or decreasing the time step by 2% in order that the iteration performance of the Poisson equation solver reaches a specified value.

3.6.5 Maximum Time Step

The user can also define the maximum time step, $\delta t_{\max imum}$, allowed for a particular calculation (default value: **deltmax** = 10^{30}). Based upon the above time-step considerations, the next computational GASFLOW-MPI time step is

$$\delta t = \min\left(\delta t_{convection}, \delta t_{diffusion}, \delta t_{combustion}, \delta t_{iteration}, \delta t_{\max imum}\right), \tag{3-116}$$

3.7 Particle Computational Model

The particle transport, deposition, and entrainment governing equations are solved independently of the fluid-flow equations. In this one-way coupled particle model, the particle/fluid volume ratio is small enough that it is assumed that the presence of the particles has no effect on the conveying gas. The particle behavior is governed by the local gas velocity field that exists after the final fluid dynamics task is performed in each computational cycle.

Each computational aerosol particle is initialized at specific physical mesh coordinates in the computational domain. The particle size and material properties are assigned to each particle. A location (x_p, y_p, z_p), velocity (u_p, v_p, w_p), diameter (d_p), and density (ρ_p) are stored for each particle.

3.7.1 Particle Equations of Motion

The fluid velocity components (u_g, v_g, w_g) at the particle location are computed from the fluid velocity components on the six faces of the cell that contains the particle. A linear interpolation is used in each direction, that is, a tri-linear interpolation. The locations of each particle are determined by

$$\begin{cases} \dfrac{dx_p}{dt} = u_p \\ \dfrac{dy_p}{dt} = \xi \dfrac{y_p}{x_p} u_p + v_p \\ \dfrac{dz_p}{dt} = w_p \end{cases} \tag{3-117}$$

where in cylindrical coordinates x_p corresponds to the radial position and y_p corresponds to the azimuthal position from a reference plane, namely the θ equal zero radians or degrees plane.

The equations of motion for each particle are as follows:

$$\begin{cases} \dfrac{du_p}{dt} = \eta \cdot \alpha_s \cdot \left(u_g - u_p\right) \\ \qquad + (1-\eta)\dfrac{3}{4} C_D \left(\dfrac{\rho_g}{\rho_p}\right) \dfrac{\left(u_g - u_p\right)\left|\mathbf{u}_g - \mathbf{u}_p\right|}{d_p} + \xi \dfrac{\left(v_p\right)^2}{r} + g_x \\ \dfrac{dv_p}{dt} = \eta \cdot \alpha_s \cdot \left(v_g - v'\right) \\ \qquad + (1-\eta)\dfrac{3}{4} C_D \left(\dfrac{\rho_g}{\rho_p}\right) \dfrac{\left(v_g - v_p\right)\left|\mathbf{v}_g - \mathbf{v}_p\right|}{d_p} - \xi \dfrac{u_p v_p}{r} + g_y \\ \dfrac{dw_p}{dt} = \eta \cdot \alpha_s \cdot \left(w_g - w'\right) \\ \qquad + (1-\eta)\dfrac{3}{4} C_D \left(\dfrac{\rho_g}{\rho_p}\right) \dfrac{\left(w_g - w_p\right)\left|\mathbf{w}_g - \mathbf{w}_p\right|}{d_p} + g_z \end{cases} , \tag{3-118}$$

where C_D, the particle drag coefficient, is defined

$$C_D = \begin{cases} \dfrac{24}{\mathrm{Re}_p}\left(1 + \dfrac{\mathrm{Re}_p^{\,2/3}}{6}\right); & \mathrm{Re}_p < 906.291 \\ 0.44 & ;\ otherwise \end{cases} \tag{3-119}$$

and Re_p, the particle Reynolds number, is

$$\mathrm{Re}_p = \frac{d_p \left|\mathbf{U}_g - \mathbf{U}_p\right|}{\nu} \quad . \tag{3-120}$$

The term $\dfrac{d}{dt}$ is the Lagrangian time derivative along the trajectory of the particle, and ρ_g is the conveying gas density. The first term on the right of each equation is the acceleration resulting from fluid drag forces. The second term in the first two equations accounts for inertial accelerations in a cylindrical coordinate system. The third term in each equation is the acceleration caused by body forces. Note that ξ is zero for Cartesian coordinates and one for cylindrical coordinates. We introduce the Stokes coefficient, α_s, which is formally

$$\alpha_s = \frac{3\pi\mu_f d_p}{m_p} \quad (3\text{-}121)$$

for $\mathrm{Re}_p << 1$. The particle drag coefficient reduces to $C_d = \frac{24}{\mathrm{Re}_p}$ for this case so that the first term is naturally contained within the second term in Equation (3-122). However it is convenient for testing purposes to allow α_s to be any positive constant and specify

$$\eta = \begin{cases} 1 \; ; \; \alpha_s > 0 \\ 0 \; ; \; otherwise \end{cases}$$

3.7.2 Particle Diffusion

Turbulent particle diffusion is modeled by a stochastic scheme developed by Hotchkiss and Hirt (Ref. 3-9). The discrete particle is considered to be a point source that diffuses in all directions and forms a conceptual cloud with a Gaussian mass distribution in each direction. The particle is to remain a point, so the Gaussian function is interpreted as a probability distribution function for the moving particle. A new location for the particle is selected using a random number generator; the probability corresponds to the mass distribution within the conceptual cloud. In this way, a Monte Carlo approximation is generated for the local diffusion of each particle. A velocity is computed from the new location and added to the conveying gas velocity corresponding to the location of that particle.

Three random numbers are selected for each particle to compute shifts in each of the three coordinate directions. The random number generator used in the code is based on a uniform distribution in the interval (0,1). This number, *x*, is transformed to a uniform distribution in the interval (-1,1) by the arithmetic operation 2*x*-1; an inverse error function then must be computed to get the desired Gaussian distribution. For computer time efficiency, we use a table of inverse error functions with 101 entries covering the error function argument from 0 to WMAX, which is an input number that has a default value of 2.0. Experience has shown this approach to be sufficiently accurate for most applications. In the particle diffusion subroutine PARTDIFF, this table is interpolated for the inverse error function for each random number, which requires much less time than the more accurate system routine for inverse error functions. A call is made to a system clock at the beginning of subroutine RPARTS, which initializes the random number generator. This call ensures a new random number sequence with each new calculation.

3.7.3 Deposition/Rebound

When particle velocity components have been computed, the particle is moved in each of the coordinate directions. Particles hitting a solid boundary are specularly reflected or deposited on the surface. A class of particles may all adhere, they may all bounce, or their behavior may be determined by the deposition/rebound model. The value of the input parameter IPDEP makes this determination. If the behavior of the particle is to be determined by the deposition/rebound model, the subroutine PARTBNC determines the particle threshold bounce velocity, $\mathbf{U}_i^*$, above which the particle may rebound:

$$\mathbf{U}_i^* = \left[\frac{2E}{m}\left(\frac{1-e^2}{e^2}\right)\right]^{\frac{1}{2}}, \tag{3-122}$$

where

$$E = \left(\frac{Ad}{12z_0}\right)\left[1+\left(\frac{A^2K^2d}{72z_0^7}\right)+\left(\frac{A^4K^4d^2}{72^2 z_0^{14}}\right)\right]+\left(\frac{4\sqrt{2d}}{15K}\right)\left(\frac{A^2K^2d}{72z_0^6}\right)^{\frac{5}{2}}. \tag{3-123}$$

The critical rebound velocity, $\mathbf{U}_i^*$, is the value of velocity for a 50% probability of bounce. The incident velocity window outside of which the particle either adheres or bounces is somewhat arbitrary, but plus or minus 50% of $\mathbf{U}_i^*$ is a reasonable assumption for this; that is, for $\mathbf{U}_i < \mathbf{U}_i^*(0.5)$ the particle always adheres, and for $\mathbf{U}_i > \mathbf{U}_i^*(1.5)$ the particle always bounces. When the velocity is in the range $0.5\,\mathbf{U}_i^* < \mathbf{U}_i < 1.5\mathbf{U}_i^*$ the determination of whether or not specific particles will bounce or adhere is made by generating a random number, α, with a value between 0 and 1, and testing for α as follows:

$$\alpha \le \left[\frac{\left(-0.5\mathbf{U}_i^* + \mathbf{U}_i\right)}{\mathbf{U}_i^*}\right]. \tag{3-124}$$

When this condition is met, the particle bounces.

The experimentally observed trend is for the coefficient of restitution, ε, to reach a maximum value at the threshold bounce velocity, and almost immediately the ratio of rebound velocity to incident velocity begins to decrease as the incident velocity increases. The following equation is used to model this behavior:

$$e = e_0\left(0.60\right)^{\exp}, \tag{3-125}$$

where

$$\exp = \frac{\mathbf{U}_i - \mathbf{U}_i^*}{10.0\mathbf{U}_i^*}, \tag{3-126}$$

and e_0 is the coefficient of restitution, the input parameter CORE, at the threshold bounce velocity.

Deposition is, in a real sense, a stochastic process that follows the general trend of the theoretical and empirical models developed and compared with available experimental data. Because of this, it is a reasonable assumption that some small, unknown percentage of the particles that impact a surface will adhere. To account for this, a percentage of all particles that impact a surface does adhere. The percentage is chosen by the input parameter DEPPERC, which has a default value of 5%. The specific particles that adhere are randomly chosen, using the random number generator.

3.7.4 Entrainment

A single particle deposited on a surface will begin to move when the forces acting in the direction parallel to the surface are zero. These opposing forces are typically the fluid-drag force and the frictional force, which is the product of the normal forces and the coefficient of friction. For a horizontal surface, the particle forces normal to the surface are adhesion, gravitational, buoyant and lift forces; for vertical surfaces the gravitational and buoyant forces are tangential to the surface. A force balance equation of these surface and aerodynamic forces is iteratively solved by a Newton-Ralphson method to determine the minimum pickup velocity of each particle. This equation is

$$C_D \mathbf{U}_{gcp}^2 = f_s \left[\frac{4}{3} d_p \frac{(\rho_p - \rho_g)}{\rho_g} \mathbf{g} - \frac{A d_p}{12 z_0^2} \left(1 + \frac{A^2 K^2 d_p}{108 z_0^7} \right) \frac{1}{0.3927 \rho_g d_p^2} + 4.11 \sqrt{\nu} \sqrt{\frac{\partial \mathbf{U}_g}{\partial y}} \mathbf{U}_{gcp} \right], \qquad (3\text{-}127)$$

where the drag coefficient is

$$C_D = 0.4 + \frac{24}{R_e} + \frac{6}{\left(1 + \sqrt{R_e}\right)}$$

and the particle Reynolds number is

$$\mathrm{Re}_{cp} = \frac{d_p \mathbf{U}_{gcp}}{\nu} \quad .$$

The vector $\mathbf{U}_{gcp}$ is the velocity of the gas at the center of the particle and is the computed particle threshold gas pickup velocity, $\mathbf{U}_{gpu0}$. Particle suspension is initiated when the velocity of the fluid flowing around the particle equals or exceeds this particle threshold suspension velocity.

This model uses a force balance approach modified by the experimental data of Cabrejos and Klinzing.* After the threshold suspension velocity in each coordinate direction for each particle is computed, these velocities are adjusted by the experimental data to obtain a semi-empirical threshold velocity. The correlation is

$$\mathbf{U}_{gpu} = \left(1.27 Ar^{-\frac{1}{3}} + 0.036 Ar^{\frac{1}{3}} + 0.45 \right) \left(0.70 Ar^{-\frac{1}{5}} + 1.0 \right) \mathbf{U}_{gpu0} \quad , \qquad (3\text{-}128)$$

---

* Francisco J. Cabrejos and George E. Klinzing, "Incipient Motion of Solid Particles in Horizontal Pneumatic Conveying," Dept. of Mech. Engr., University of Pittsburg, unpublished paper (1991).

where the Archimedes number is

$$Ar = \frac{\mathbf{g}}{\nu^2}\frac{\left(\rho_p - \rho_g\right)}{\rho_g} d_p^3 \quad .$$

The semi-empirical threshold velocity component is computed for each particle for each coordinate direction.

The semi-empirical threshold velocity is used to test for particle entrainment. The orientation of the solid boundary on which the particle is located in the computational mesh has been determined, and the information is stored in the array MPAC for each deposited particle. In addition, the velocity in each coordinate direction at the particle location has been computed and is stored. The determination of these velocities uses the law-of-the-wall equation, as discussed in Section 2.9.3.1, to estimate these velocities at the particles, which are typically embedded in the viscous, inner sublayer of the boundary layer. Again considering the stochastic behavior of particles, a probability of entrainment is computed. This is not based on experimental data but is solely an artifice to broaden the critical velocity at which particles will be suspended. When $0.8\mathbf{U}_{gpu} \le \mathbf{U}_{gcp} < \mathbf{U}_{gpu}$, the probability of entrainment is

$$PRB_{ntrn} = 4.0\left(\frac{\mathbf{U}_{gcp} - 0.8\mathbf{U}_{gpu}}{\mathbf{U}_{gpu}}\right). \tag{3-129}$$

When $\mathbf{U}_{gcp} \ge \mathbf{U}_{gpu}$,

$$PRB_{ntrn} = 0.8 + \left(\frac{\mathbf{U}_{gcp} - \mathbf{U}_{gpu}}{\mathbf{U}_{gpu}}\right) \quad . \tag{3-130}$$

This gives an 80% probability of entrainment when the gas velocity at the particle location, $\mathbf{U}_{gcp}$, is equal to the semi-empirical threshold velocity, $\mathbf{U}_{gpu}$, and a 100% probability of entrainment when the velocity at the particle location is equal to or greater than 1.2 times the semi-empirical threshold velocity. The particle is never entrained when the gas velocity at the particle location is less than or equal to 0.8 times the semi-empirical threshold velocity. A random number, α, is generated and used to determine if the particle is actually entrained. When $PRB_{ntrn} \ge \alpha$, the particle is entrained. In this case, the particle is assigned the local fluid velocity components parallel to the surface. The particle velocity component normal to the surface is set to zero. The particle is also moved to a point directly out from (normal to) the surface where the specific particle was located. The distance from the surface at which the particle is set is between one-half and one computational cell dimension. The exact location in this range is randomly chosen.

3.7.5 Two-way momentum coupling

In previous GASFLOW version (GASFLOW 3.1 for instance), the Lagrangian discrete particle dynamic model is based on uncoupled approach or one-way fluid and particle coupling, which means the particle motion is affected by the fluid phase but not vise versa. This assumption is often made because it is easier to compute the fluid and particles motion separately if the effect of particles on the fluid is negligible.

In case of high particle volume fraction or mass loading, one-way fluid particle coupling is not sufficient. When the volume fraction of the dispersed phase can not be neglected compared to the continuous phase, the interaction between the continuous and dispersed materials can not be neglected. In this case, the effect of the discrete phase on the continuum must be considered, such as the mass, momentum, turbulence and energy couplings.

In GASFLOW 3.2, the model for momentum exchange between fluid and particles are developed. To have an computationally effective, stable algorithm for GASFLOW-MPI, an implicit two-way fluid particle coupling method within the framework of the GASFLOW-MPI code is developed. It should be noted that the volume fraction of the continuous phase should be far greater than the volume fraction of the dispersed phase so that the pressure gradient is negilible to the fluid acceleration.

Considering the effect of particles, the momentum equation of the continuous phase can be written as

$$\frac{\partial(\rho_g u_g)}{\partial t} + \nabla \cdot \left(\rho_g \vec{U}_g \vec{U}_g\right) = -\nabla \mathrm{P} + \nabla \cdot \vec{\tau} + \rho_g \vec{g} + S_{mom,p} + S_{mom,others} \tag{3-131}$$

where $S_{mom,p}$ is the source term which take into account the contribution of each individual particle to the momentum change of the continuous phase. $S_{mom,others}$ is other sources which may change the momentum of the continuous phase.

3.7.6 Numerical Solution Method

The basic procedure for advancing the particle transport solution through one increment of time, Δt, consists of four steps for each particle.

(1) Compute the fluid velocity at the particle location. This local fluid velocity, which is used to calculate the fluid drag force, has two parts: the velocity interpolated from the computed velocity field (Section 3.4) and a diffusion velocity computed from a Monte-Carlo method (Section 3.7.2).

(2) Compute intermediate particle velocities from explicit approximations of the momentum equations [Equation 3-121] without fluid drag forces.

(3) Compute a new time-level particle velocity by iteratively adjusting the intermediate particle velocities to include the effects of the particle fluid drag forces.

(4) For two-way momentum coupling, integrate the particle drag forces in the fluid cells and put it as source term into the momentum equation of the continous phase.

(5) Move the particles to a new location using the average particle velocity for the time increment, *Δt*.

Following the transport of the particles, two additional phenomena are modeled.

(1) Deposit particles impacting a rigid surface, if criteria discussed in Section 3.7.3 are met.

(2) Entrain particles deposited on surfaces, if criteria discussed in Section 3.7.4 are met.

Here we only deal with the Cartesian geometry case involving relative velocities. The equation of particle motion can be rewritten as

$$\begin{cases} F_x(u_p) = \left(u_p\right)^{n+1} - \left(u_p\right)^n \\ \qquad -\eta\cdot\alpha_s\cdot\left(u_f - u'\right)^{n+1}\cdot\delta t - (1-\eta)\cdot\beta^{n+1}\cdot\left(u_f - u'\right)^{n+1}\cdot\delta t \\ F_y(v_p) = \left(v_p\right)^{n+1} - \left(v_p\right)^n \\ \qquad -\eta\cdot\alpha_s\cdot\left(v_f - v'\right)^{n+1}\cdot\delta t - (1-\eta)\cdot\beta^{n+1}\cdot\left(v_f - v'\right)^{n+1}\cdot\delta t \\ F_z(w_p) = \left(w_p\right)^{n+1} - \left(w_p\right)^n \\ \qquad -\eta\cdot\alpha_s\cdot\left(w_f - w'\right)^{n+1}\cdot\delta t - (1-\eta)\cdot\beta^{n+1}\cdot\left(w_f - w'\right)^{n+1}\cdot\delta t \end{cases} \tag{3-132}$$

where $\mathbf{u}' = \mathbf{u}_p + \mathbf{u}_{diff}$.

The particle diffusion velocity, $\mathbf{u}_{diff}$, represents a measure of the turbulence due to turbulent fluctuations on the mean flow, and hence influence on particle motion. We assume isotropic locally homogeneous turbulence, so we define

$$\mathbf{u}_{diff} = \pm\sqrt{\frac{4\lambda}{\delta t}}\,erf^{-1}(\zeta) \tag{3-133}$$

where λ is the particle turbulent diffusion or dispersion coefficient, δt the particle time step for the numerical scheme, and ζ a random number between zero and one. The sign, $\pm$, is also determined randomly.

There are many models for $\mathbf{u}_{diff}$ in the literature. For the time being, we will use Equation 3-137 and provide (1) a constant value for λ or (2) a functional value for λ that makes use of local turbulence properties of the flow field. We see that

$$\beta = \frac{3}{4}C_d\left(\frac{\rho_g}{\rho_p}\right)\frac{\left|\mathbf{u}_g - \mathbf{u}'\right|}{d_p} \tag{3-134}$$

and

$$\left|\mathbf{u}_f-\mathbf{u}'\right|=\left[\left(u_f-u'\right)^2+\left(v_f-v'\right)^2+\left(w_f-w'\right)^2\right]^{1/2} \tag{3-135}$$

Our task is to find the advanced time levels for each velocity component, the superscript (n+1) values. It's enough then to force the left hand side of Equation 3-136 to zero. We employee a Newton-Raphson numerical iteration scheme to accomplish this task. The algorithm successively iterates u_p, v_p, and w_p until the left hand side is within a small tolerance of zero.

Since each component of Equation 3-136 is similar, we develop only the x component or u_p equation. For this purpose, we say

$$\overline{\overline{u}}_p=\overline{u}_p-\frac{F_x(\overline{u}_p)}{\frac{dF_x(\overline{u}_p)}{d\overline{u}_p}} \tag{3-136}$$

with

$$F_x(\overline{u}_p)=\left(\overline{u}_p\right)^{n+1}-\left(\overline{u}_p\right)^{n}-\eta\cdot\alpha_s\cdot\left(u_{eff}-\overline{u}_p\right)^{n+1}\cdot\delta t-(1-\eta)\cdot\beta^{n+1}\cdot\left(u_{eff}-\overline{u}_p\right)^{n+1}\cdot\delta t \tag{3-137}$$

$$\frac{dF_x(\overline{u}_p)}{d\overline{u}_p}=1+\eta\cdot\alpha_s\cdot\delta t+(1-\eta)\cdot\beta^{n+1}\cdot\delta t\cdot\left(1+\frac{\left(u_{eff}-\overline{u}_p\right)^2}{\left|\mathbf{u}_{eff}-\overline{\mathbf{u}}_p\right|^2}\right) \tag{3-138}$$

where $\mathbf{u}_{eff}=\mathbf{u}_f-\mathbf{u}_{diff}$ and $\beta=\frac{3}{4}C_d\left(\frac{\rho_f}{\rho_p}\right)\frac{\left|\mathbf{u}_{eff}-\overline{\mathbf{u}}_p\right|}{d_p}$.

To simulation two-way momentum coupling, we need to calculate the momentum source term for the continuous phase. For the particles (total number of particles, *np*) which are located in the fluid cell, *m*, the volumetric particles contributions to the momentum of the fluid cell are computed as,

$$S_{mom,p,x}^{n+1}(m)=\sum_{i=1}^{np}\left\{-\frac{m_p(i)}{Vol_g(m)}\frac{3}{4}C_d(i)\frac{\rho_g(m)}{\rho_p(i)}\frac{\left|\mathbf{u}_g^{n+1}(i)-\mathbf{u}_p^{n+1}(i)\right|}{d_p(i)}\left[u_g^{n+1}(i)-u_p^{n+1}(i)\right]\right\} \tag{3-139}$$

$u_g^{n+1}(i)$ is the local fluid velocity at the exact location of particle, *i*, which are obtained by interpolation of the relevant fluid velocities. Vol_g is the volume of the fluid cell.

References

Ref. 3-1. C. W. Hirt, A. A. Amsden, and J. L. Cook, "An Arbitrary Lagrangian-Eulerian Computing Method for All Flow Speeds," J. Comp. Phys. **14**, 227 (1974).

Ref. 3-2. A. A. Amsden and C. W. Hirt, "YAQUI: An Arbitrary Lagrangian-Eulerian Computer Program for Fluid Flow at All Speeds," Los Alamos Scientific Laboratory report LA-5100 (March 1973).

Ref. 3-3. J. D. Ramshaw and J. K. Dukowicz, "APACHE: A Generalized-Mesh Eulerian Computer Code for Multicomponent Chemically Reactive Fluid Flow," Los Alamos Scientific Laboratory report LA-7427 (January 1979).

Ref. 3-4. A. A. Amsden, H. M. Ruppel, and C. W. Hirt, "SALE: A Simplified ALE Computer Program for Fluid Flow at All Speeds," Los Alamos Scientific Laboratory report LA-8095 (June 1980).

Ref. 3-5. L. D. Cloutman, J. K. Dukowicz, J. D. Ramshaw, and A. A. Amsden, "CONCHAS-SPRAY: A Computer Code for Reactive Flows with Fuel Sprays," Los Alamos National Laboratory report LA-9294-MS (May 1982).

Ref. 3-6. A. A. Amsden, J. D. Ramshaw, P. J. O'Rourke, and J. K. Dukowicz, "KIVA: A Computer Program for Two- and Three-Dimensional Fluid Flows with Chemical Reactions and Fuel Sprays," Los Alamos National Laboratory report LA-10245-MS (February 1985).

Ref. 3-7. A. A. Amsden, P. J. O'Rourke, and T. D. Butler, "KIVA-II: A Computer Program for Chemically Reactive Flows with Sprays," Los Alamos National Laboratory report LA-11560-MS (1989).

Ref. 3-8. B. van Leer, "Towards the Ultimate Conservation Difference Scheme IV. A New approach to Numerical Convection," J. Comp. Phys., 23, 276-299 (1977).

Ref. 3-9. R. S. Hotchkiss and C. W. Hirt, "Particulate Transport in Highly Distorted Three-Dimensional Flow Fields," Proceedings of the 1972 Summer Computer Simulation Conference, Vol. 2, pp. 1037-1041 (AAIA, San Diego, CA, 1972).

4 Summary

In this report we have described the theory behind the parallel GASFLOW-MPI computer code and the computational model used to numerically integrate the governing equations in time and space.

The core of the code is based on the conservation laws for mass, momentum, and energy, which were presented in integral form to more closely mirror the finite volume formulation for the discrete equations. The basic fluid dynamics equations are supplemented with relations for multiple gas species, heat transfer, structural condensation and vaporization of liquid films, a two-phase homogeneous equilibrium model for the fluid mixture, a parametric droplet depletion or rainout model, turbulence transport, a finite-rate global chemical kinetics model for combustion, an ignitor model for both spark- and glow-plug-type ignitors, recombiner box models for the NIS and Siemens devices, and a discrete particle model for transport, deposition, and entrainment.

An ICE'd-ALE numerical method, adapted to fixed geometry, is used to solve the equations of motion. Each time step is split into three phases: (1) an explicit Lagrangian phase in which most of the physics and chemistry is evaluated, (2) an implicit Lagrangian phase in which time-advanced pressure, velocity, density, and energy are computed together, and (3) a rezone phase in which problem variables are repartitioned or rezoned onto the original mesh. Turbulence quantities and a computational time-step stability analysis then complete the time step.

Using the above field equation model coupled with finite-rate global chemical kinetics, we have successfully analyzed the hydrogen transport, mixing, and combustion and mitigating measures with recombiners and ignitors in a nuclear reactor containment under accident conditions. The reader is referred to the GASFLOW-MPI Assessment and Verification Manual for more details.